13
Topics in Heterocyclic Chemistry

Series Editor: R. R. Gupta

Topics in Heterocyclic Chemistry

Series Editor: R. R. Gupta

Recently Published and Forthcoming Volumes

Bioactive Heterocycles VI
Flavonoids and Anthocyanins in Plants, and latest Bioactive Heterocycles I
Volume Editor: N. Motohashi
Volume 15, 2008

Heterocyclic Polymethine Dyes
Synthesis, Properties and Applications
Volume Editor: L. Strekowski
Volume 14, 2008

Synthesis of Heterocycles via Cycloadditions II
Volume Editor: A. Hassner
Volume 13, 2008

Synthesis of Heterocycles via Cycloadditions I
Volume Editor: A. Hassner
Volume 12, 2008

Bioactive Heterocycles V
Volume Editor: M. T. H. Khan
Volume 11, 2007

Bioactive Heterocycles IV
Volume Editor: M. T. H. Khan
Volume 10, 2007

Bioactive Heterocycles III
Volume Editor: M. T. H. Khan
Volume 9, 2007

Bioactive Heterocycles II
Volume Editor: S. Eguchi
Volume 8, 2007

Heterocycles from Carbohydrate Precursors
Volume Editor: E. S. H. El Ashry
Volume 7, 2007

Bioactive Heterocycles I
Volume Editor: S. Eguchi
Volume 6, 2006

Marine Natural Products
Volume Editor: H. Kiyota
Volume 5, 2006

QSAR and Molecular Modeling Studies in Heterocyclic Drugs II
Volume Editor: S. P. Gupta
Volume 4, 2006

QSAR and Molecular Modeling Studies in Heterocyclic Drugs I
Volume Editor: S. P. Gupta
Volume 3, 2006

Heterocyclic Antitumor Antibiotics
Volume Editor: M. Lee
Volume 2, 2006

Microwave-Assisted Synthesis of Heterocycles
Volume Editors: E. Van der Eycken, C. O. Kappe
Volume 1, 2006

Synthesis of Heterocycles via Cycloadditions II

Volume Editor: Alfred Hassner

With contributions by
K. G. Abhilash · V. Nair · K. V. Radhakrishnan · K. M. L. Rai
G. Rajendar · R. S. Rao · B. V. S. Reddy · J. S. Yadav

Springer

The series *Topics in Heterocyclic Chemistry* presents critical reviews on "Heterocyclic Compounds" within topic-related volumes dealing with all aspects such as synthesis, reaction mechanisms, structure complexity, properties, reactivity, stability, fundamental and theoretical studies, biology, biomedical studies, pharmacological aspects, applications in material sciences, etc. Metabolism will be also included which will provide information useful in designing pharmacologically active agents. Pathways involving destruction of heterocyclic rings will also be dealt with so that synthesis of specifically functionalized non-heterocyclic molecules can be designed.
The overall scope is to cover topics dealing with most of the areas of current trends in heterocyclic chemistry which will suit to a larger heterocyclic community.
As a rule contributions are specially commissioned. The editors and publishers will, however, always be pleased to receive suggestions and supplementary information. Papers are accepted for *Topics in Heterocyclic Chemistry* in English.
In references *Topics in Heterocyclic Chemistry* is abbreviated *Top Heterocycl Chem* and is cited as a journal.

Springer WWW home page: springer.com
Visit the THC content at springerlink.com

ISBN 978-3-642-09710-2 e-ISBN 978-3-540-78373-2
DOI 10.1007/978-3-540-78373-2

Topics in Heterocyclic Chemistry ISSN 1861-9282

Softcover reprint of the hardcover 1st edition 2008

Cover design: WMXDesign GmbH, Heidelberg

Printed on acid-free paper

9 8 7 6 5 4 3 2 1 0

springer.com

Topics in Heterocyclic Chemistry Also Available Electronically

For all customers who have a standing order to Topics in Heterocyclic Chemistry, we offer the electronic version via SpringerLink free of charge. Please contact your librarian who can receive a password or free access to the full articles by registering at:

springerlink.com

If you do not have a subscription, you can still view the tables of contents of the volumes and the abstract of each article by going to the SpringerLink Homepage, clicking on "Browse by Online Libraries", then "Chemical Sciences", and finally choose Topics in Heterocyclic Chemistry.

You will find information about the

- Editorial Board
- Aims and Scope
- Instructions for Authors
- Sample Contribution

at springer.com using the search function.

Color figures are published in full color within the electronic version on SpringerLink.

Preface

Heterocyclic molecules play a significant role in life processes and have played a major role in industrial developments of the last century, for instance in the field of dyes, pharmaceuticals, pesticides, polymers etc. They comprise not only some of the most interesting and biologically important natural products like alkaloids, carbohydrates, nucleic acids, and antibiotics but include many practical drugs and a large segment of known synthetic organic compounds. Hence scientists have devoted a great amount of effort to find optimal synthetic approaches to a variety of heterocyclic compounds.

Among the most successful and selective synthetic processes are cycloaddition reactions, since they involve simultaneous or sequential formation of two or more bonds often with a high degree of stereoselectivity and regioselectivity. For instance, 1,3 dipolar cycloadditions, which are electronically equivalent to Diels–Alder reactions, are among the most-common 5-membered ring-forming systems. In addition they usually proceed with a high degree of stereo- and regio-control. It is therefore, not surprising that synthesis of many important classes of heterocycles, including those of useful biologically active molecules, have utilized cycloaddition steps in their formation. Furthermore, many heterocycles serve as intermediates in the synthesis of polyfunctional molecules.

Volume I of "Synthesis of Heterocycles via Cycloadditions" featured five chapters on the following topics:

- Isoxazolines from Nitro Compounds: Synthesis and Applications;
- Cycloaddition Reactions of Azides Including Bioconjugation;
- Enantioselective Cycloadditions of Azomethine Ylides;
- Heterocycles by Cycloadditions of Carbonyl Ylides Generated from Diazo Ketones;
- Heterocycles from Unsaturated Phosphorus Ylides.

In this volume we present four selected contributions by well-known authors, each an authority in his field. The first chapter is devoted to the use of oximes in cycloadditions which leads to formation of isoxazolines and isoxazolidines and from there to synthesis of macrolides like amphotericin and of other natural products and bioactive molecules. Furthermore, 4+2 cycloadditions of nitrosoalkenes are also included. This chapter complements the one in the previous volume which discussed access to isoxazolines via nitroalkanes.

The subject of the second chapter is how pyrylium and pyridinium betaines can be used in cycloadditions leading to interesting N- and O-bridged heterocycles and applications to synthesis of a variety of natural products. Reactions such as 6+3 cycloadditions are also included.

The third chapter deals with synthesis of heterocycles via cycloadditions catalyzed by indium derivatives and related Lewis acids. Metal-catalyzed cycloadditions play an important role in generation of heterocyclic compounds. In particular catalysis by indium salts, which can tolerate the presence of water, can be used in the synthesis of many types of heterocycles such as 3-membered ring aziridines, 4-membered rings by 2+2 cycloadditions, as well as various 5-membered ring heterocycles and 6-membered heterocyclic systems via hetero-Diels–Alder reactions.

The last chapter brings to light the formation of heterocyclic rings via higher-order cycloadditions such as 8+2 and 6+4 annulation reactions. In this manner bicyclic systems including azaazulenes can be constructed. I want to thank all authors for their excellent presentations and their splendid cooperation.

This volume is dedicated with love to my grandchildren Ariel, Amit, Matan, Tal, Hadas, and Tamar.

Ramat Gan, February 2008 Alfred Hassner

Contents

Contents of Volume 12

Synthesis of Heterocycles via Cycloadditions I

Volume Editor: Hassner, A.
ISBN: 978-3-540-78368-8

Top Heterocycl Chem (2008) 13: 1–69
DOI 10.1007/7081_2008_112

Published online: 4 March 2008

Heterocycles via Oxime Cycloadditions

Kuriya Madavu Lokanatha Rai

Department of Chemistry, University of Mysore, Manasagangotri, 570 006 Mysore, India
kmlrai@yahoo.com

Abstract Isoxazolines and isoxazols serve as important building blocks in the construction of new molecular systems and functional groups. Syntheses of these building blocks were easily achieved by the cycloaddition of nitrile oxides derived from oximes with alkenes. This chapter comprises the utilization of oximes for the synthesis of various biologically active molecules such as ptilocaulin, amphotericin, macrolides, etc. Synthesis of oxazines via 4+2 cycloaddition of in situ generated nitrosoolefins from ketoximes with alkenes are also included in this chapter.

Keywords Chloramine-T · Intramolecular nitrile oxide cycloaddition (INOC) · Intramolecular oxime-olefin cycloaddition (IOOC) · Isoxazolines · Isoxazols · Nitrile oxide · Nitrosoolefin · Oxazines · Oxime

Abbreviations

Bn	Benzyl
CAN	Ceric ammonium nitrate
CAT	Chloramine-T
DDQ	Dichlorodicyanoquinone
DMF	Dimethylformamide
DIPT	Diisopropyl tartrate
DIP-Cl	β-Chloroisopinocamphenyl borane
FMO	Frontier molecular orbital
HOMO	Highest occupied molecular orbital
INOC	Intramolecular nitrile oxide cycloaddition

IOOC	Intramolecular oxime-olefin cycloaddition
LUMO	Lowest unoccupied molecular orbital
NCS	*N*-chlorosuccinimide
PEG 400	Polyethylene glycol 400
PTBIB	Protein tyrosine phosphatase inhibitors
QCM	Quadruple cycloaddition
Ts	Tosyl=(*p*-toluenesulfonyl)
UDP	Uridine diphosphate

1 Introduction

Cycloaddition reactions comprise a process in which two or more π systems combine to form a stable cyclic molecule, during which sigma bonds are formed between the termini of π systems and no fragment is lost. A concerted mechanism requires a single transition state and no intermediate, which lies on the reaction path between reactants and adduct. The two important cycloaddition reactions that usually occur by the concerted mechanism and involve a 4π and a 2π component are (1) the Diels–Alder reaction, and (2) 1,3-Dipolar cycloaddition reaction.

1.1 1,3-Dipolar Cycloaddition Reactions

The general concept of 1,3-dipolar cycloaddition reactions evolved out of the monumental work carried out in the early 1960s by Huisgen and his coworkers. In these reactions, a 5-membered ring is formed by the cycloaddition of a three-atom entity, a 4π component called a 1,3-dipole and a two-atom entity, a 2π component called a dipolarophile. In all 1,3-dipoles, there are four electrons in three overlapping π-orbitals. From the resonance structures contributing to the dipole, it is clear that the 1,3-dipoles can be both nucleophilic and electrophilic in nature. This ambivalence of the 1,3-dipole is of key importance in understanding its reactivity. The nucleophilic character of the 1,3-dipole may be stronger than its electrophilic quality. The stereochemistry of a 1,3-dipolar cycloaddition reaction is a stereospecific *syn* addition with respect to dipolarophile.

The most general approach to the synthesis of 5-membered heterocycles involves cycloaddition of a 1,3-dipole to an appropriate unsaturated substrate, the dipolarophile. Intermolecular cycloaddition results in the formation of one new ring only. When the 1,3-dipole and the substrate are part of the same molecule, cycloaddition is intramolecular and leads to a new bicyclic system. The intramolecular 1,3-dipolar cycloaddition is a powerful method for the construction of fused ring heterocycles and has been applied to benzene rings *ortho* substituted with 1,3-dipole and dipolarophile functions. In particular intramolecular nitrile oxide cycloaddition results in dihydroisoxazole (isoxazoline) derivatives, which are precursors for γ-amino alcohols, β-hydroxy ketones and derivatives, useful in the synthesis of natural products (Eq. 1).

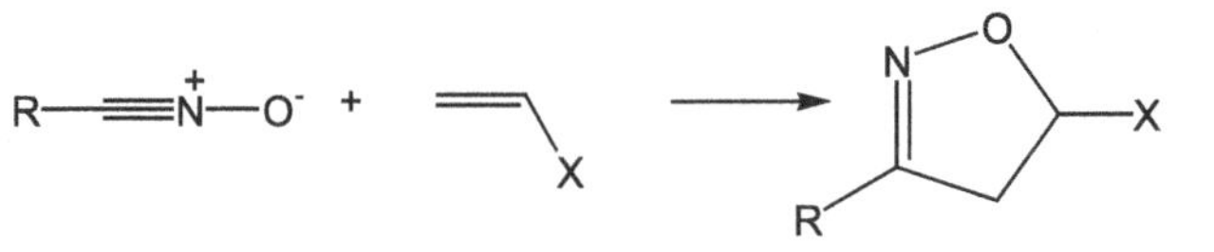

(1)

There are exhaustive review articles on this topic, available in the literature [1–15]. Therefore, only the important recent reactions of nitrile oxides are outlined here.

1.2 Generation of Nitrile Oxides

All known methods for the synthesis of nitrile oxides start with an organic framework already containing the C–N–O sequence necessary for conversion to the nitrile oxide ($R{-}C{\equiv}N^{+}{-}O^{-}$) structure. Many methods are reported to generate nitrile oxide. Common methods of generating nitrile oxides involve the oxidative dehydrogenation of aldoximes [16], the dehydration of primary nitro compounds with aryl isocyanate [17–19] or other reagents [18, 19], and the dehydrohalogenation of hydroxyiminoyl halides [20, 21]. Hydroximoyl chlorides are generated from oximes by chlorination with chlorine, *N*-chlorosuccinimide, nitrosyl chloride, sodium hypochlorite or *tert*-butyl hypochlorite [22]. Tokunaga et al. utilized silver acetate for the generation of nitrile oxide starting from hydroximoyl halides [23]. Loupy et al. developed a new method for the generation of nitrile oxides by microwave irradiation of hydroximoyl chlorides in the presence of dipolarophiles [24]. Nitrile oxides are generated from *O*-trimethylsilylhydroximoyl chlorides by treatment with potassium fluoride in acetonitrile at ~20 °C or from hydroximoyl chlorides using molecular sieves (3–5 Å) in CH_2Cl_2 [14].

Nitrile oxides are often unstable at room temperature and need to be formed in situ in the presence of a dipolarophile. Several oxidative dehydrogenation methods of aldoximes using oxidants such as lead tetraacetate, alkali hypohalite [25], *N*-bromosuccinimide in dimethyl formamide followed by base treatment [26], 1-chlorobenzotriazole [27], chloramine-T [28] mercuric acetate [29] are reported. In situ generation of nitrile oxide from aldoxime by potassium ferric cyanide requires aqueous medium [30], while ceric ammonium nitrate (CAN) can be used only for aromatic aldoximes [31, 32]. Radhakrishna et al. [33] reported the use of hypervalent iodine compounds as an oxidizing agent for the in situ conversion of aldoximes to nitrile oxides.

Since the workup requires alkaline conditions, this method is limited to alkaline resistant compounds [33]. Moreya et al. reported the in situ generation of nitrile oxides by the reaction of aldoximes with tertiary butyl hypochlorite and bis(tributyltin) oxide [34]. The reaction proceeded efficiently under mild condition in which *O*-stannylated aldoximes are thought to be the intermediate (Scheme 1).

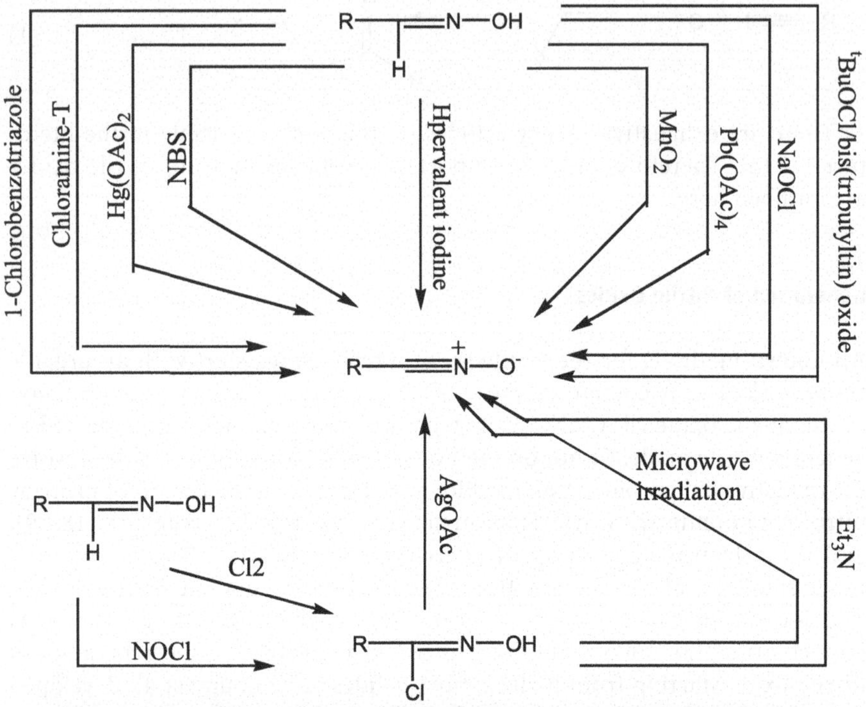

Scheme 1

The method by Rai and Hassner [28], using chloramine-T as dehydrogenating reagent, not only allows in situ generation but also often allows the isolation of nitrile oxides from aldoximes. The in situ reaction is usually carried out by heating a mixture of aldoxime and an alkene in ethanol in the presence of chloramine-T. By employing this method, it was possible to isolate and characterize the nitrile oxide, of which some are liquids and some are solids. The unstable nitrile oxide, identified by NMR spectrometry, slowly dimerized on standing alone, or in the presence of added vinyl sulfone it underwent cycloaddition to yield an isoxazoline in good yield [35, 36].

Manganese(IV) oxide (MnO_2) was found to be effective for the in situ generation of nitrile oxides from aldoximes. Keigiel et al. proposed the following mechanism for the oxidation of aldoxime to nitrile oxide (path B), analo-

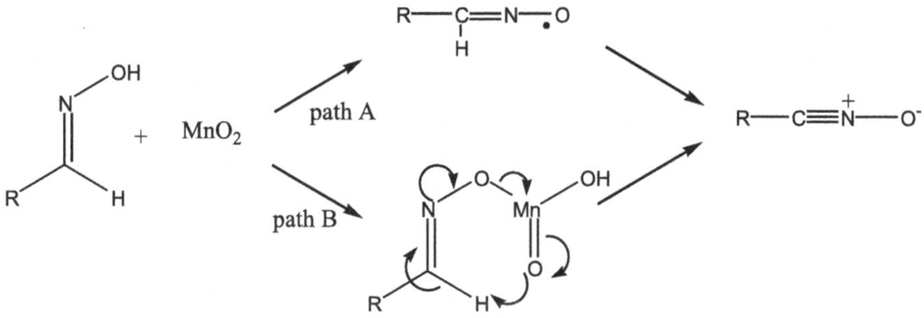

Scheme 2

gous to the oxidation of aldoximes to nitrile oxide by lead tetraacetate [37] (Scheme 2).

1.3 Reactions of Nitrile Oxides

Nitrile oxides readily undergo 1,3-dipolar cycloaddition reactions with various dipolarophiles. Alkenes and alkynes serve as excellent dipolarophiles. Cycloaddition of nitrile oxides to olefins yields isoxazolines, while addition of nitrile oxide to alkynes yields isoxazoles directly. If the dipolarophile possesses more than one set of unsaturation as in an enyne, addition to either (or both) site(s) may occur. Indeed with nitrile oxides as the dipole and a 1,3-enyne as substrate, the chemoselectivity is very sensitive to the substitution pattern of the enyne, so that either product (**1**) or (**2**) may predominate (Scheme 3) [38].

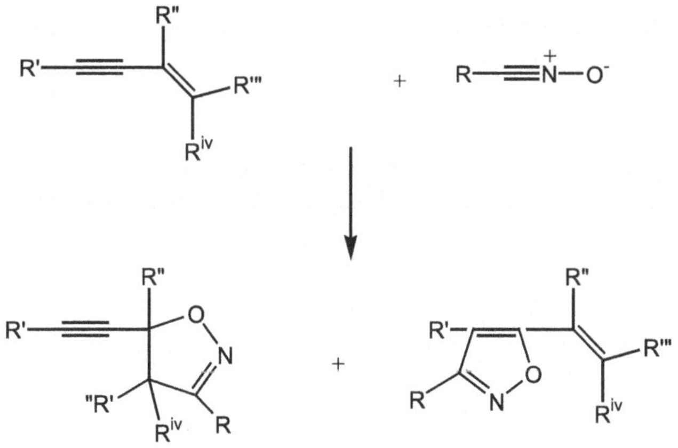

1 **2**

Scheme 3

Unlike the frequently non-selective reaction of 1,3-enynes with a 1,3-dipole, nitrile oxides add chemo, regio and stereoselectively to the free double bond of (1,3-enyne)$Co(CO)_6$ complexes to provide 5-alkenyl-2-oxazole derivatives (see **2** in Scheme 3) in moderate yield [39].

The ability to add nitrile oxides is not restricted to C–C multiple bonds. Thus a C=O group may act as a dipolarophile and yield 1,3,4-dioxazoles. However the C=O group is less reactive than a C=C function as a dipolarophile. This is clearly shown by the reaction of nitrile oxide with acetylacetone. Here acetylacetone prefers to react as an enol (80%) rather than as a ketone (Scheme 4) leading to two regioisomers [40].

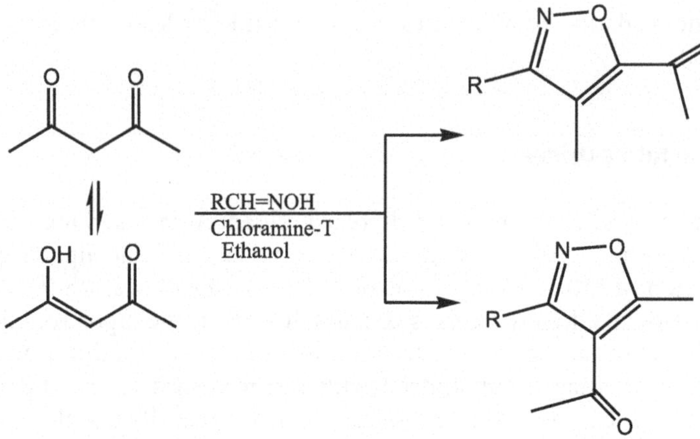

Scheme 4

Though the C=S group is not a good dipolarophile in Diels–Alder reactions, it is very reactive in 1,3-dipolar cycloadditions of nitrile oxides and yields 1,4,2-oxathiazolines. The C≡N group normally does not undergo 1,3-dipolar cycloaddition reaction because it is a poorer dipolarophile compared to the C=C group. Thus, in the case of acrylonitrile, nitrile oxide reacts with alkene to form cyano-substituted 2-isoxazoline. However, if the C=C bond is deactivated by multiple substitution, the cyano-group may become a better dipolarophile (Scheme 5) Thus, tetracyanoethylene adds a nitrile oxide yielding a 1,2,4-oxadiazole derivative as one of the products. Apart from this dipolarophile, nitrile oxides add to other multiple bonds to give corresponding 5-membered heterocycles.

Shang et al. [41] synthesized 1,2,4-oxadiazoline derivatives for the first time on soluble polymer support via 1,3-dipolar cycloaddition of nitrile oxides to an imine (Scheme 6). This one pot protocol affords the required oxadiazolines in good to high yield with excellent purity (>91%) after cleavage from a commercially available PEG-4000 as support.

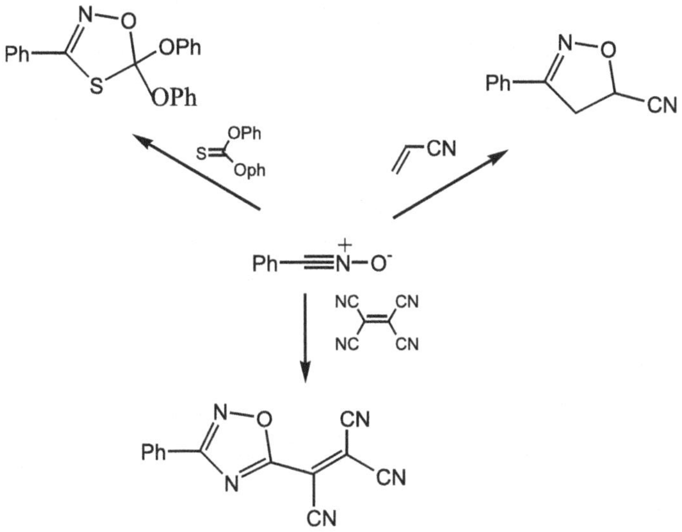

Scheme 5

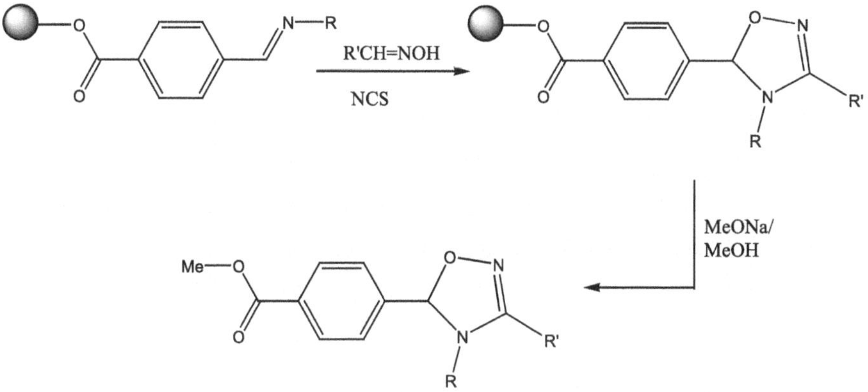

Scheme 6

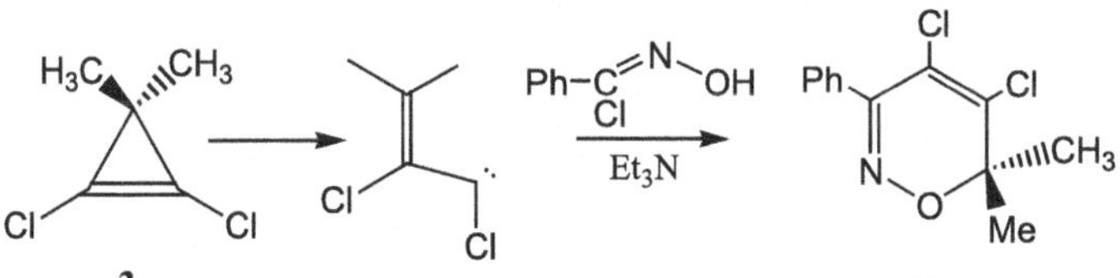

3 **4** **5**

Scheme 7

The unusual formal [3+3] cycloaddition of a nitrile oxide with vinyl carbene (**4**) derived by the ring opening of the cyclopropene (**3**) yields 1,2-oxazines (**5**) in moderate to good yield but it is not clear whether the 1,2-oxazine is directly formed by concerted cycloaddition or by a stepwise process (Scheme 7) [42].

1.4
Orbital Symmetry Analysis of 1,3-Dipolar Cycloaddition Reactions: FMO Method

In a 1,3-dipolar cycloaddition reaction, two new sigma bonds are formed at the expense of π-electrons of the reactants. The concerted reaction results from the overlap of orbitals of one molecule (dipolar) with the orbitals of the other (dipolarophile). As in the case of electrocyclic reactions, here also one can concentrate on the HOMO. If one does so, what is seen is that the HOMO of each reactant is fully occupied by two electrons. Therefore, HOMO–HOMO interaction is not possible. In order to form a bond, each HOMO has to overlap with an empty orbital, namely with the most stable of the empty orbitals, i.e., the LUMO.

In the transition state of 1,3-dipolar cycloadditions, stabilization chiefly comes from the overlap between the HOMO of one reactant (dipole or di-

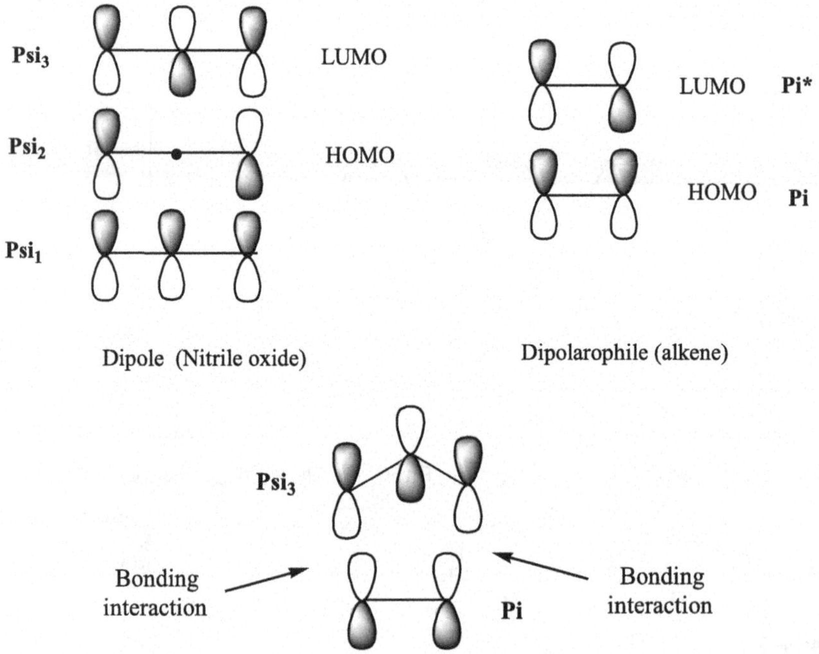

polarophile) with the LUMO of the other (dipolarophile or dipole) in bonding fashion.

Let us first consider the LUMO of the 1,3-dipole and the HOMO of the dipolarophile.

Let us now consider the HOMO of the dipole and the LUMO of the dipolarophile.

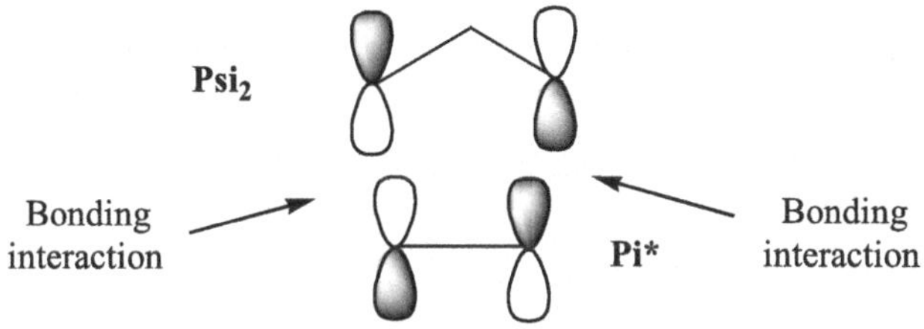

Thus 1,3-dipolar cycloaddition is a thermally allowed reaction. It also follows that the nature of substituents on the double bond will influence the regiochemistry of the cycloaddition, namely formation of 4- or 5-substituted isoxazolines (see below).

For photochemical reactions, the HOMO in the excited state of one reactant (dipole/dipolarophile) and the LUMO in the ground state of another (dipolarophile/dipole) have to be considered. HOMO of the excited state of a 1,3-dipole is ψ_3 and LUMO of the ground state of the dipolarophile is π^*.

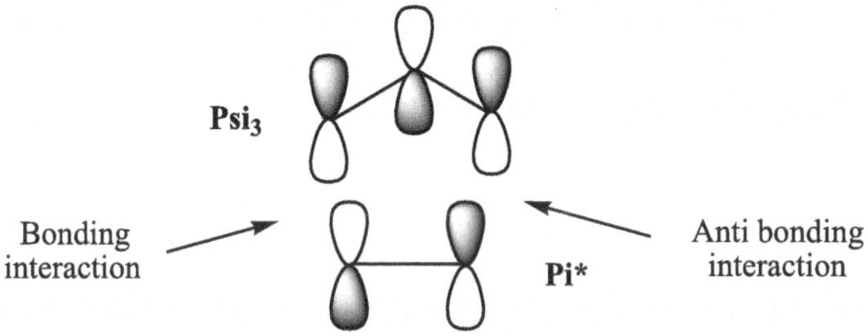

Here, on one side there is an antibonding situation, hence no product is formed. Therefore, $[\pi_s^4 + \pi_s^2]$ photochemical reactions are forbidden.

1.5 Stereo- and Regiochemistry of 1,3-Dipolar Cycloaddition Reactions

The stereochemistry of 1,3-dipolar cycloaddition reaction is a stereospecific *syn* addition with respect to dienophile, meaning that the stereochemistry of the double bond is preserved in the product. With some dipoles, two possible diastereomers can be formed by *syn* addition due to two differing orientations of the reactant molecules. The possibility of two orientations for addition determines the regiochemistry of the isoxazoline (Scheme 8).

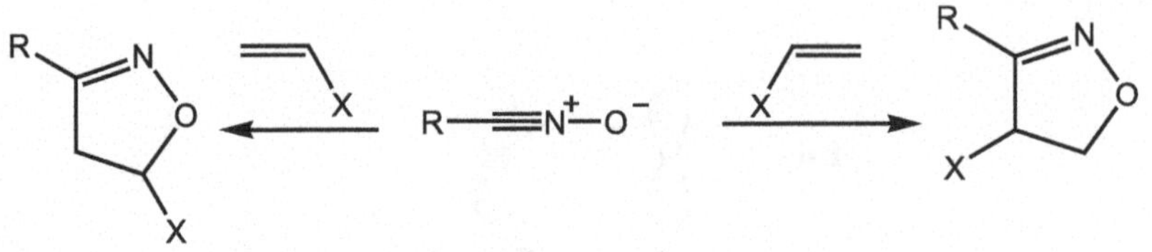

Scheme 8

The regioselectivity can be interpreted in terms of interaction between the FMO of 1,3-dipole and dienophile. Usually, for dipolarophiles with electron-attracting groups, the dipole–HOMO and dipolarophile–LUMO interaction is dominant. The reverse is true for dipolarophiles with electron-donating groups. However, there are HOMO–LUMO interactions of comparable magnitude.

According to the principle of maximum overlap, the preferred isomers of each interaction can be predicted by union of two sites of the reactants having the largest coefficient value [43]. Most dipolarophile undergo cycloaddition to give 5-substituted isoxazolines with high stereoselectivity. As the electron affinity of the dipolarophile increases, an increasing tendency towards production of the 4-substituted isoxazoline is found. At some point, there must be a switch over from LUMO to HOMO control as one increases the electron withdrawing power of the substituent on the alkene.

1.6 Application of Nitrile Oxide Cycloaddition Reactions

1,3-Dipolar cycloaddition of a nitrile oxide to the C=C bond of a dipolarophile is of considerable importance in organic synthesis, since this reaction yields isoxazoles, or 2-isoxazolines which serve as important building blocks in the construction of new molecular systems and functional groups. First, these heterocycles can be very efficiently prepared from readily available precursors, e.g., aldehyde oximes. Secondly, they can be conveniently modified, thus allowing the transformation of a molecule with simple structure to functionally complex derivatives. Thirdly, a suitable pattern of substituents makes the isoxazoline ring survive under a variety of chemical reaction conditions, allowing manipulation in other parts of the molecule. Finally, the lability of the nitrogen–oxygen bond to catalytic or chemical reduction under mild conditions unravels a vast array of differ-

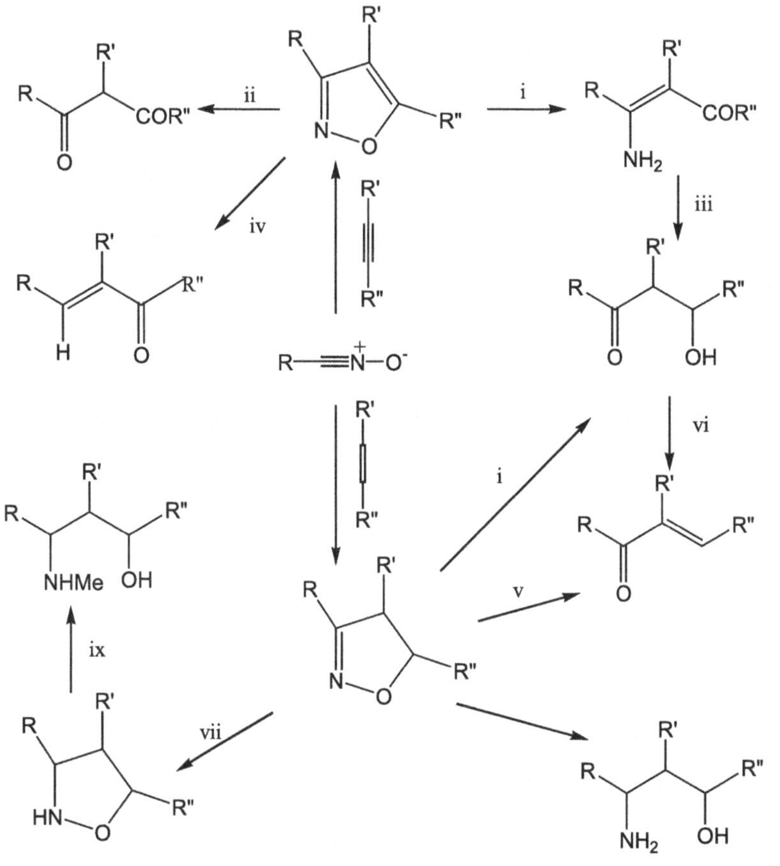

i. H_2, catalyst; or $Mo(CO)_6$ ii. H_3O^+

iii. PhCOCl, Py, $NaBH_4$, AcOH;

iv. Birch reduction, TsOH

v. LDA in THF, $TiCl_3$/HCl

vi. NaOH; vii. $NaBH_4$, MeOH;

viii. $LiAlH_4$; ix. MeI, H_2 catalyst or ZnOH

Scheme 9

ent functionalities (Scheme 9). For instance, Stork and coworkers [44] have used the isoxazole group as a masked ketone in the synthesis of cyclohexenones and steroids. Baraldi et al. [45] synthesized β-hydroxy ketones from isoxazolines utilizing molybdenum hexacarbonyl as catalyst for the ring cleavage.

1.6.1
Intermolecular Cycloaddition Reactions of Nitrile Oxides

Stereoselective 1,3-dipolar addition of bromonitrile oxide to S(+)-isopropylidine-3-butene-1,2-diol represents the key step in the preparation of a potent muscarinic receptor [46], while 1,3-dipolar cycloaddition of nitrile oxide derived from hydroximoyl chloride to an alkene possessing a lactone moiety provides a useful alternate route to the synthesis of phyllanthocin, an aglycone of the glycoside phyllanthoside isolated from the roots of the South American tree *Phyllanthus acuminatus* [47].

Cycloaddition of in situ generated nitrile oxide obtained via an oxime from (dialkoxyphosphoryl) carbonyl halide with 1-bromo-3,3-dimethyl cyclopropane leads to an isoxazole (Scheme 10), while reaction with 3,3-dimethyl-1,2-dichlorocyclopropane leads to an oxazine derivative (see Scheme 7) [41]. Formation of the isoxazole in Scheme 10 may be explained by the opening of the cyclopropane ring in the adduct with bromine migration or via the ring opening of the cyclopropane with formation of 3-bromo-3-methyl-1-butyne followed by the cycloaddition of nitrile oxide to this -yne.

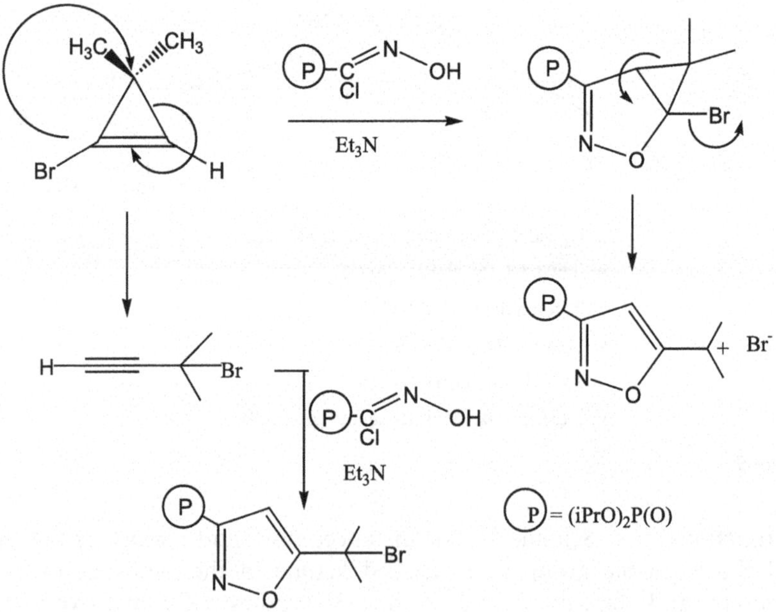

Scheme 10

Fiddouli et al. [48] successfully reported one pot diastereoselective synthesis of chiral 1,4,2-oxathiazoline by 1,3-dipolar cycloaddition of a nitrile oxide

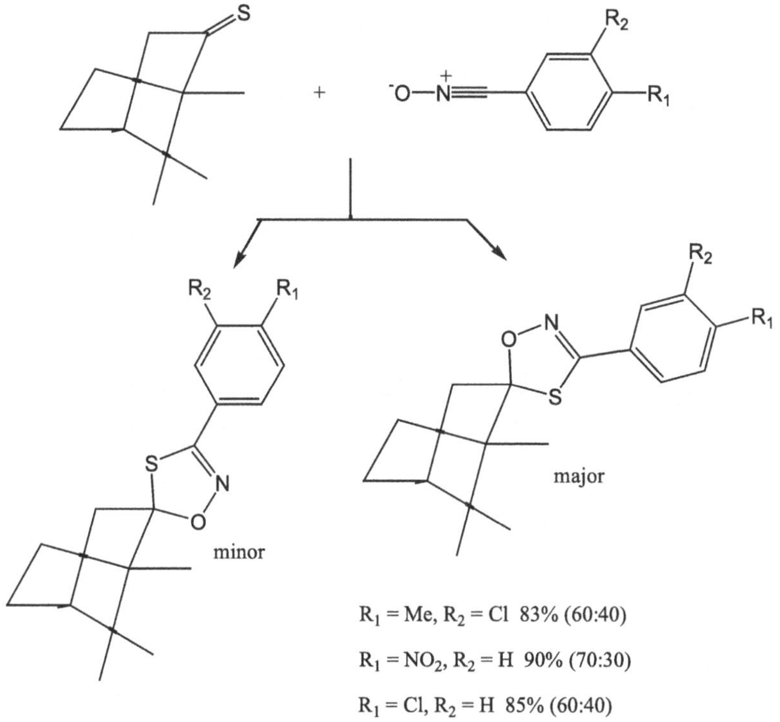

Scheme 11

generated from arylhydroximoyl chloride with (1*R*)-thiocamphor. In this reaction, C=S behaves as the dipolarophile (Scheme 11).

Dehaen et al. [49] were able to derivatize artemisinin, a potent antimalarial drug isolated from *Artemisia annua*, employing a nitrile oxide cycloaddition strategy. They observed low diastereoselectivity during the cycloaddition (Scheme 12).

Zerressen et al. [50] reported the isolation of RNA molecules catalyzing a 1,3-dipolar cycloaddition between a nitrile oxide and an acrylate conjugated to RNA. This strategy involved the construction of a randomized pool of RNA conjugates containing the dipolarophile substrate acrylate attached to the RNA by ligation of the functionalized dinucleotide moiety. The functional group was linked to the dinucleoptide via a flexible ethylene glycol containing a photosensitive group which could ideally be utilized to eliminate RNA conjugates catalyzing undesired reactions to other positions of the RNA chain. The conjugated RNA pool was then incubated with an in situ generated biotinylated nitrile oxide to a select RNA molecule that catalyzes the simultaneous formation of C–O and C–C bond, resulting in a heterocyclic isoxazoline ring. The obtained RNAs would be linked to biotin and could

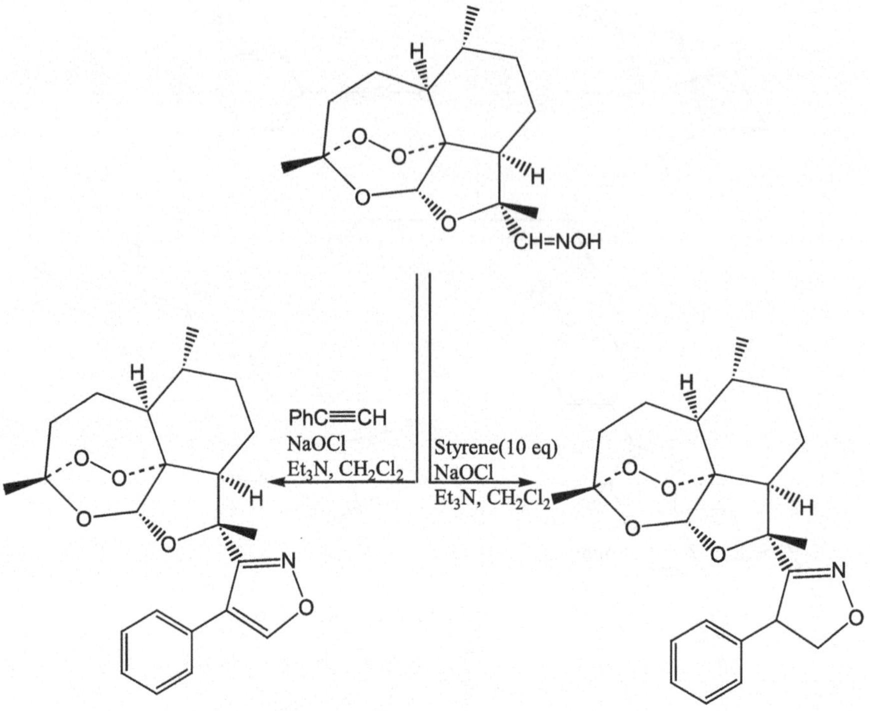

Scheme 12

therefore be isolated by affinity chromatography on a streptavidin matrix (Scheme 13).

Furan and thiophene annelated β-hydroxy chalcones were synthesized via 1,3-dipolar cycloaddition reactions. The scheme involves the cycloaddition of a nitrile oxide with styrene followed by mild oxidation with DDQ and subsequent catalytic hydrogenation using Raney nickel as catalyst (Eq. 2) [51].

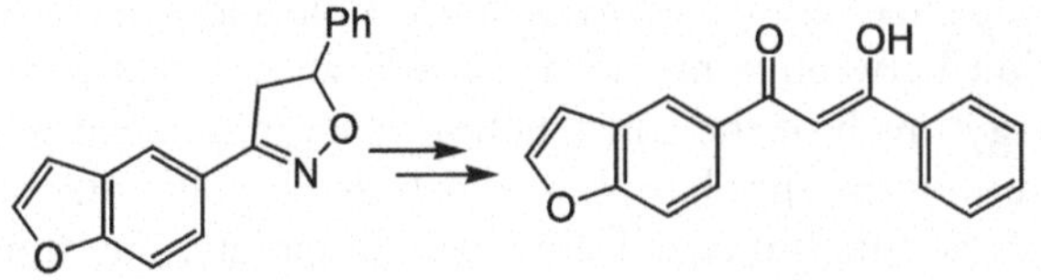

(2)

The oximes (**7**) derived from naturally occurring furanoflavonoids (**6**) were subjected to oxime-olefin cycloaddition reaction with substituted olefins (styrene, 4-methylstyrene, 2-vinylpyridine, 4-vinylpyridine, 1-vinylimidazole) in the presence of chloramine-T to produce isoxazolines (**8**) in good yield (Scheme 14) [52]. All the synthesized compounds were evaluated against PTPase enzymes. Compound (**8f**) and (**8g**) displayed significant inhibitory

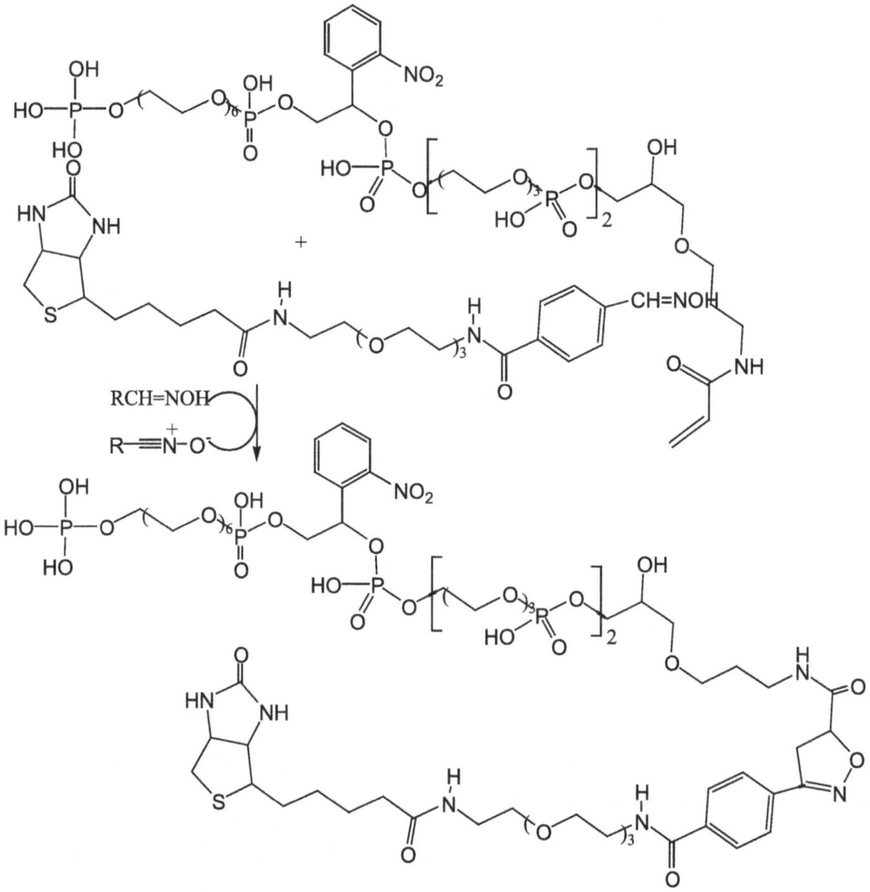

Scheme 13

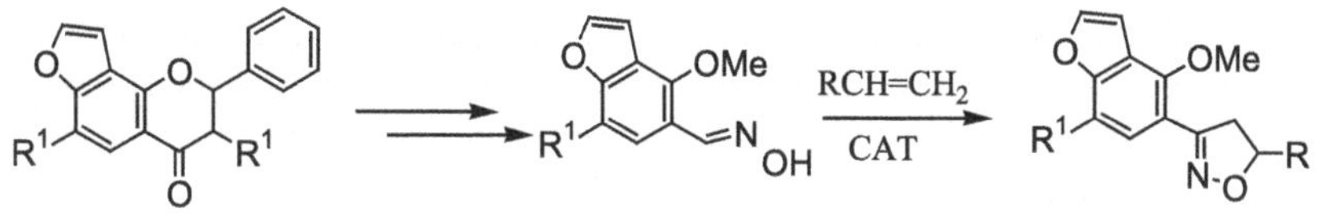

6 7 8

Scheme 14

activity with IC_{50} values 76 and 81 μM, respectively (Table 1). The structure-activity relationship study shows that substitution at C-5 of isoxazole does not cause any significantly important inactivity except for (**8b**), whereas introduction of methoxy at C-7 in compounds (**8f**) and (**8g**) remarkably enhanced the activity profile as compared to the reference compound, sodium vanadate.

Table 1 PTPIB inhibitory activity of compounds (**8**)

Product	R^1	R	Inhibition	IC_{50} (μM)	Ki(μM)
a	H	Ph	15.8	-	-
b	H	$-C_6H_4(COCH_3)$	22.5±2.69	227	45
c	H	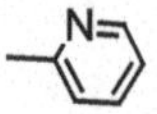	49.1±7.71	150	27.5
d	H	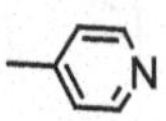	17.4±0.42	-	-
e	H	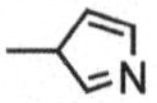	14.9±2.92	-	-
f	OCH_3	Ph	80.4±0.54	76	30
g	H	$-C_6H_4(OCH_3)$	79.6±3.31	81	323.0
h	H	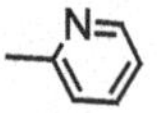	184±6.65	-	-
Na_3VO_4			56.2	-	-

IC_{50} concentration for 50% inhibition

Chai et al. [53] showed that the success of 1,3-dipolar cycloaddition reactions of ylidenepiperazinediones with mesitonitrile oxide is governed by a number of factors. The stereoselectivities of reaction with chiral ylidenepiperazine-2,5-dione can be directed by judicious choice of substituent on the *N* and/or *C* of the piperazine ring. For instance, to achieve the transition state for leading to the 4,4-disubstituted regioisomer, a significant steric repulsion between the piperazinedione ring and the bulky mesityl system would be present as shown in Scheme 15. This repulsion is minimized in the transition state leading to 5,5-disubstituted cycloadduct. It was suggested that the steric effect outweighs the electronic consideration and the transition state that leads to the 5,5-disubstituted regioisomer is favored.

The stereochemistry of the major isomer of (**9b**) formed from the reactions of the diacylated methylidinepiperazinedione was found to result from the attack of the dipole onto the face of the dipolarophile opposite (*anti*) to the remote α-methyl group. In contrast, X-ray structure analysis of a crystal of the major isomer of cycloadduct (**9a**) showed the opposite stereochemical outcome in which the attack of the dipole occurred *syn* to the remote α-methyl substituent (Scheme 15).

In addition, molecular modeling studies suggest that the piperazinedione ring of the diacetylated dipolarophile is more puckered than that of the monoacetylated methylidine piperazinedione. Then a *N,N'*-diacetylated com-

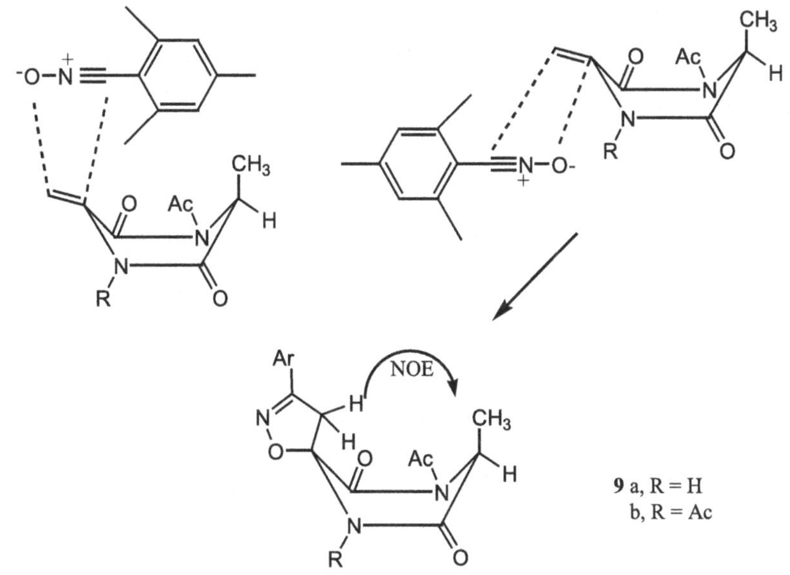

9 a, R = H
b, R = Ac

Scheme 15

pound would be predicted to show greater facial selectivity as compared to the monoacetylated piperazinedione.

Kikuchi et al. observed that inclusion of a heteroatom in the carotenoid ring reduces the toxicity 1000-fold and inclusion of a heteroatom in the ring of *trans*-retinoic acid reduces the toxicity threefold [54]. Prompted by these results, Simoni et al. [55] synthesized isoxazole carotinoids (**10**) via intermolecular nitrile oxide-olefin cycloaddition (Scheme 16). They observed that isoxazole (**10b**) represents a novel retinoid endowed with apoptotic activity in

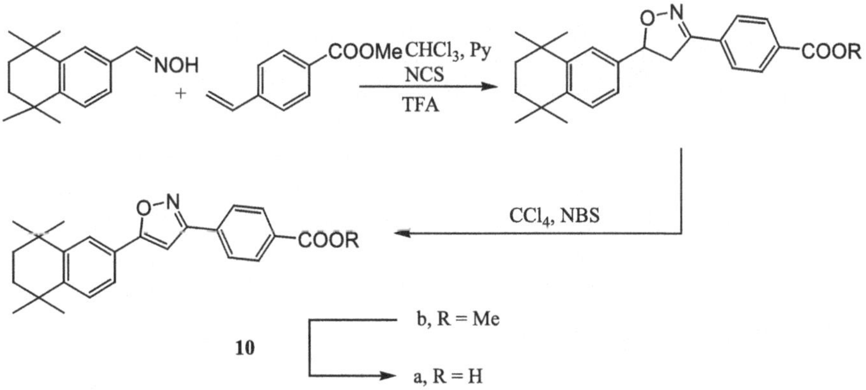

Scheme 16

MDR cells. Its ability to act in K562 and HC60R cell lines suggest that it may have important application in the treatment of different leukemias.

An effective and chiral specific synthesis of DMP 754 (**11**), a novel peptide, orally active and extremely potent platelet (**11b**) and (**11a**) antagonist involves 1,3-dipolar cycloaddition of an oxime-derived nitrile oxide to isobutyl vinyl acetate as a key step [55].

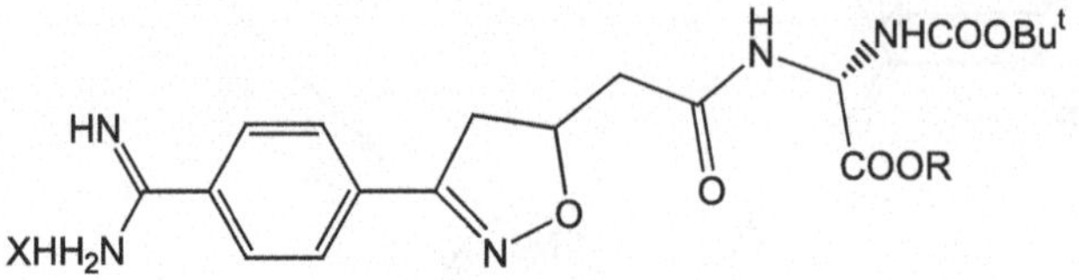

11 a: X = CH_3COO , R = CH_3
b: X = Cl, R = CH_3
c: X = $PhSO_2$, R = H

Recently Zierke [57, 58] showed that the isoxazoline derivatives of the type (**12**) act as agrochemical fungicides and plant growth regulators, especially (**12**, R′=CF_3; R = *p*-Cl–C_6H_4–) as a 0.035% spray gave 95–100% control of cucumber mildew. Coumarins (**13** and **14**) were screened for antiinflammatory activity in vivo using carrageenin rat Pew edema and in vitro through their antiproteolytic activity and their ability to inhibit β-glucuronidase and 12-lipoxygenase [59]. These compounds are attainable via oxime-olefin cycloadditions.

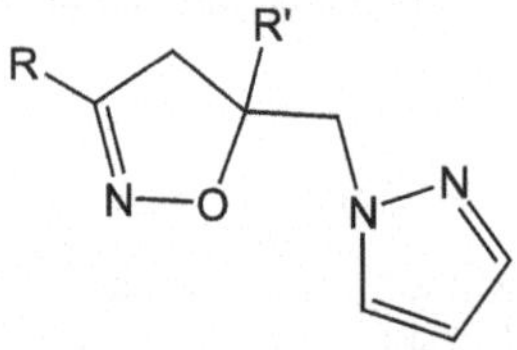

12

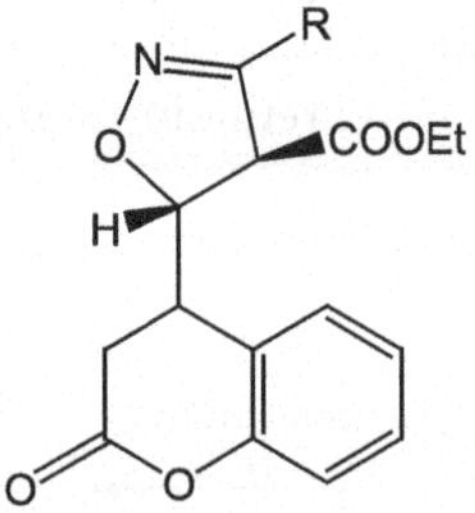

13

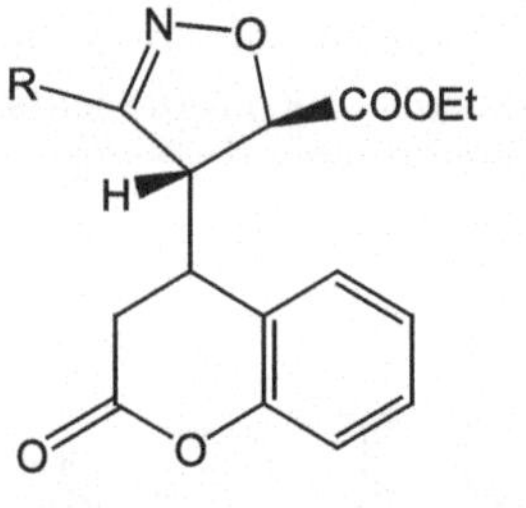

14

Cohan et al. [60, 61] found that isoxazolines (**15**), which can be prepared from oximes, are useful in the treatment of inflammatory conditions or diseases including rheumatoside arthritis, osteoarthritis, asthma, bronchitis, chronic obstruction, septic shock, tuberculosis, graft versus host disease and cachexia anovil with AIDS or cancer.

Diphenyl acetohydroximoyl chloride can be easily isolated as a pure white solid since it has low solubility in organic solvents. Hence, it can be a very useful precursor in the nitrile oxide-olefin cycloaddition. The result-

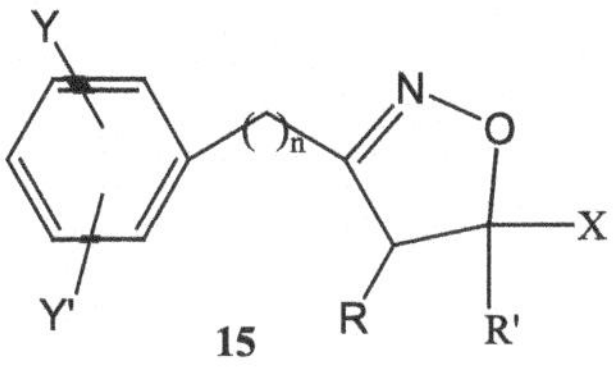

15

X =(CH2)qOH, CH(OH)R^2, (CH_2)mCONR^3OH, q,m = 0-5; R^3 = C_1-C_4 alkyl;

n =0,3; Y = Y' =H, 1-6 alkyl, CF3, halo, OR^4; R^4 = alkyl, fluroalkyl.

ing 2-isoxazoline, possessing an alkenyl moiety can undergo electrophilic iodoether formation using iodine or iodine monochloride, affording highly substituted tetrahydrofuran derivatives. In the mechanism of the iodoetheration reaction, the cation stabilizing effect of the 3-substituent (R_1) in the isoxazoline was very important, in order to promote Beckmann cleavage to form 2-cyanomethyl-3-hydroxy-5-iodomethyl tetrahydrofuran (Scheme 17) [62].

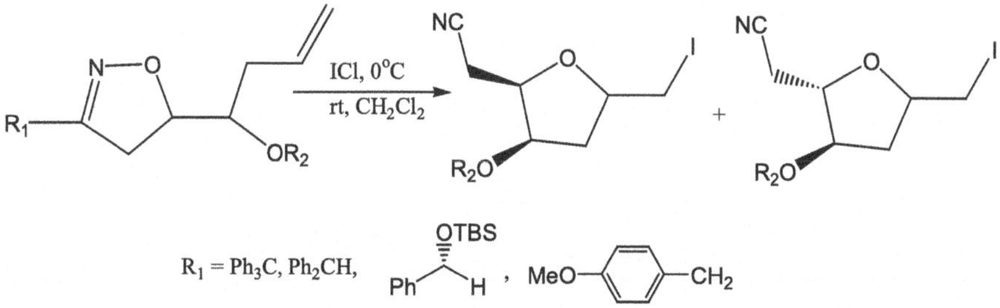

Scheme 17

Sultams, the cyclic counterpart of sulfonamide derivatives, exhibit a vast variety of biological activities. Therefore, synthesis of fused heterocycles possessing the sultam framework are of interest and this can be easily achieved by 1,3-dipolar cycloaddition strategy. Nitrile oxides synthesized from the corresponding hydroxymoyl chloride cycloadd to non-racemic α,β-unsaturated-γ-sultams (**16**) to give optically pure bicyclic isoxazolines (Scheme 18). The activation of the double bond in (**16**) by the sultam functionality is strong enough to render possible its cycloaddition reaction with nitrile oxides at room temperature. Nitrile oxides incorporating a phenyl substrate with two electron donating methoxy groups exhibited high reactivity towards (**16**) affording the adducts in very high yield [63].

The *exo*-methylenepyrrolidine systems in Scheme 19 undergo a highly regioselective 1,3-dipolar cycloaddition reaction with nitrile oxides generated from the corresponding aldoximes leading to spirooxazolinoproline-based amino acids in good yields and with ca. 1:4 *cis:trans* diastereoselectivity (Table 2) [64].

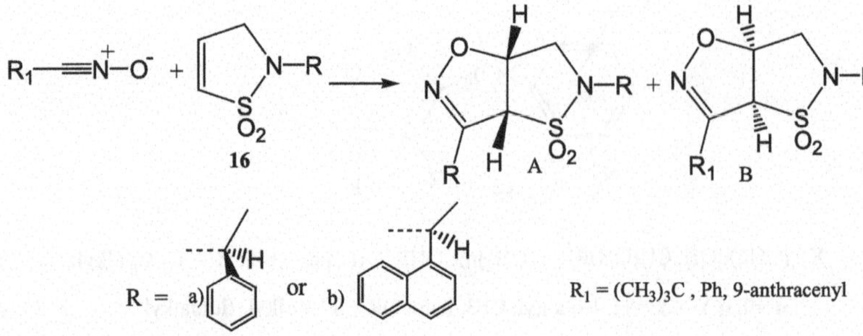

Scheme 18

Table 2 Results of 1,3-dipolar cycloaddition of nitrile oxides and chiral unsaturated sultams

Entry	Dipole	R	Time (h)	Product (dr)	Yield (%)
1	$R_1 = (CH_3)_3C$	a	36	A:B (1.3)	50
2	R_1 = Ph	a	30	A:B (1.5)	56
3	R_1 = 9-anthracenyl	a	36	A:B (1.2)	58
4	$R_1 = (CH_3)_3C$	b	36	A:B (1.3)	42
5	R_1 = Ph	b	36	A:B (1.2)	70
6	R_1 = 9-anthracenyl	b	30	A:B (1.2)	55

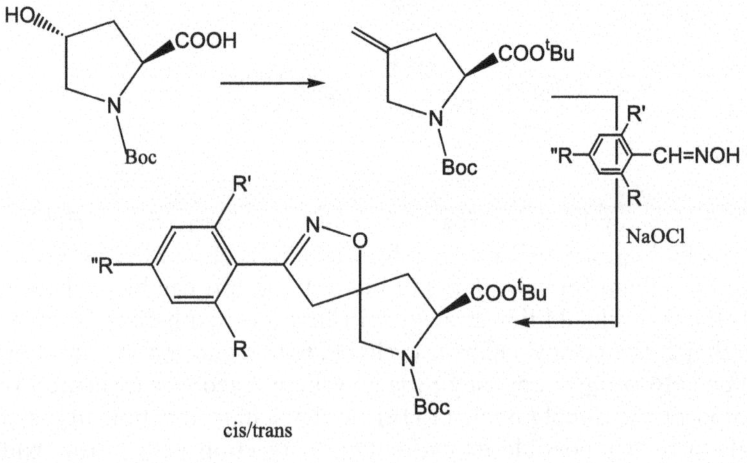

Scheme 19

	Cis:trans	Yield
R′ = R″ = H, R = OMe	20:80	75%
R = R″ = H, R′ = OMe	19:80	74%
R = R″ = Cl, R′ = H	22:78	75%

In the presence of $Zr(O^tBu)_4$, the nitrile oxide generated from ethyl chlorooximinoacetate cycloadds with a vinyl epoxide, affording a spiroisoxazoline as a mixture of two diastereomers (Scheme 20). Porco et al. used this strategy as a key step for the total synthesis of the spiroisoxazoline natural product (+)calfianin [65].

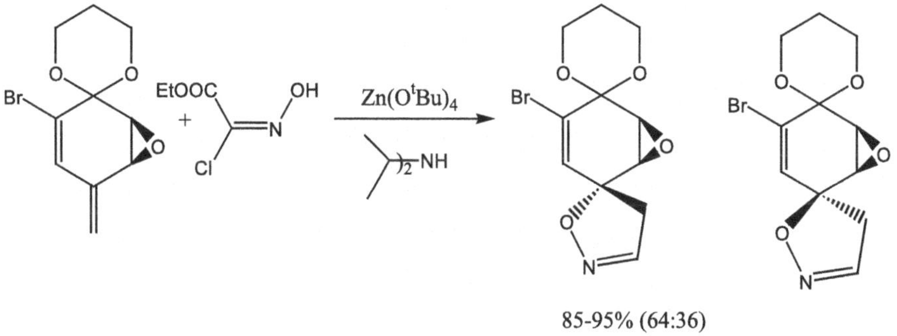

Scheme 20

UDP-3-*O*-[R-3-hydroxymyristoyl]-GlcNAc (deacetylase) (LpxC) is an amidase that catalyzes the second step of lipid biosynthesis in gram-negative bacteria. Known inhibitors of this enzyme are oxazolines incorporating a hydroxamic acid at the 4-position, which is believed to coordinate to the single essential zinc ion. Pirrung et al. showed that isoxazolines each incorporating a different potential metal binding functional group were found to exhibit enzyme inhibitory activity, including one that is more active than the corres-

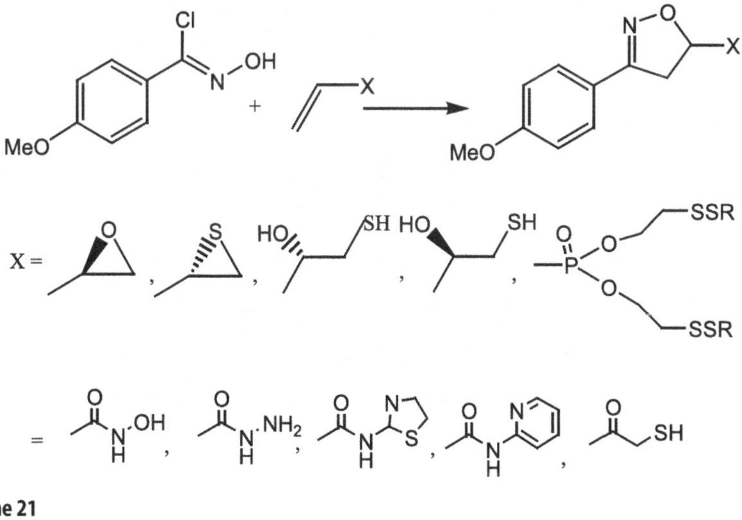

Scheme 21

ponding hydroxamic acid [66]. The active isoxazolines were prepared via an intramolecular cycloaddition of in situ generated nitrile oxide, from a hydroximoyl chloride, with various olefin bearing functional groups (Scheme 21).

Optically active oximes (**16**) and (**17**) prepared from (*R*)- and (*S*)-3-hydroxy-2-methyl-propanoic acid methyl ester react smoothly with enantiomerically pure *cis*-allylic alcohols to give the corresponding 4,5-syn cycloadducts with complete regio and stereoselectivity [67] (Schemes 22 and 23). The free al-

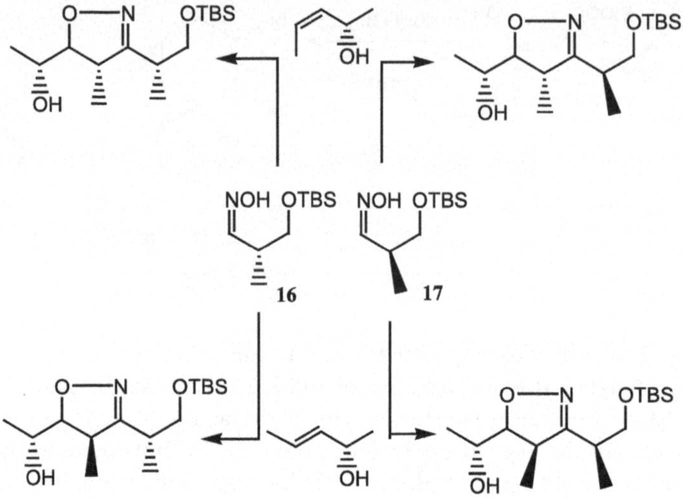

Scheme 22

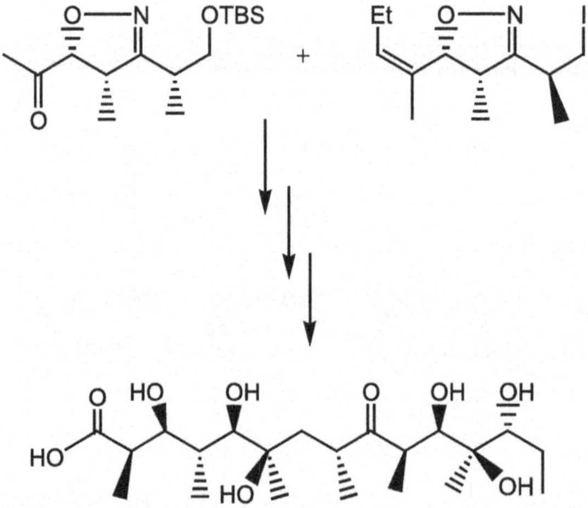

Erythronolide A sec-acid

Scheme 23

cohols present in the cycloadducts provide a convenient synthetic handle for further transformation to more elaborate structures. Using this synthon, Bode et al. [67] were able to synthesize erythronolide A seco-acid, a polyhydroxy compound.

The isoxazoline product derived from cycloaddition of the nitrile oxide derived from protected glycol aldoxime with 2-butene was cleaved by hydrolysis utilizing Raney nickel in a mixture of acetic acid, methanol and water (1:8:2) [68]. The resulting β-hydroxy ketone is a precursor for the synthesis of 2-methylene-γ-lactone, which is an important subunit of many natural products (Scheme 24).

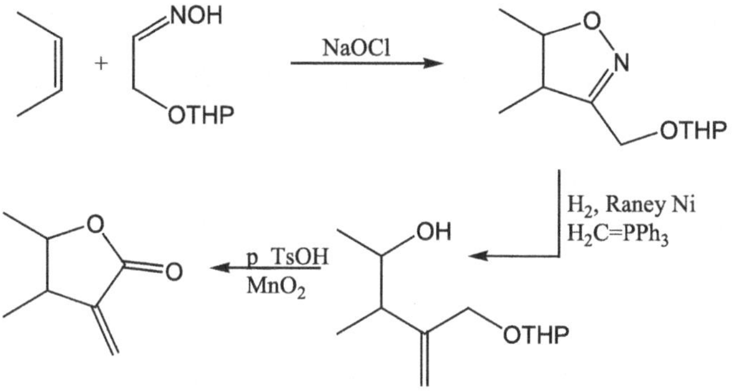

Scheme 24

Padmavathi et al. synthesized in almost quantitative yields novel keto linked bis heterocycles having two different heterocyclic rings. They utilized bischalcones [69] and bis sulphones [70–72] as dipolarophiles and nitrile oxides as 1,3-dipoles (Scheme 25).

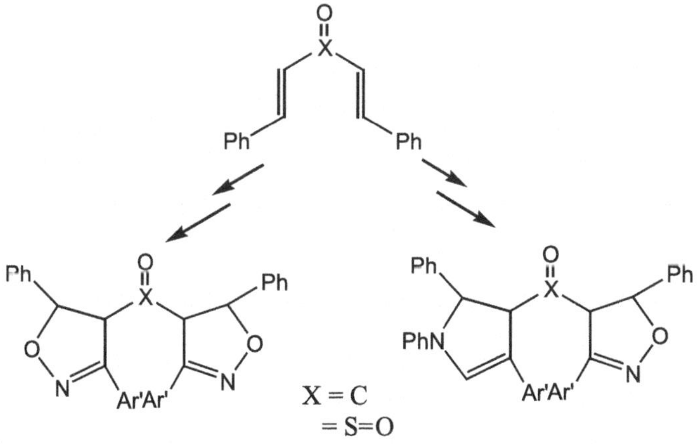

Scheme 25

Ukaji et al. [73] observed that the use (*R*,*R*)-diisopropyl tartrate as chiral auxilary in the presence of diethyl zinc for the cycloaddition of a nitrile oxide to an achiral allylic alcohol leads to the formation of optically active 2-isoxazolines in 95% ee (Scheme 26). The stereochemical course of the cycloaddition was found to be similar to the asymmetric Simmons–Smith reaction, that is, the nitrile oxides will add from the front side of the alcohol.

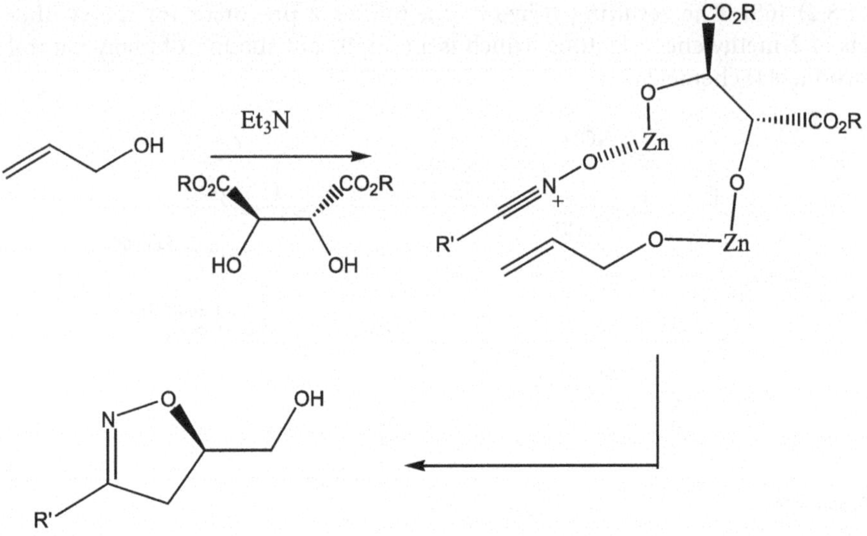

Scheme 26

Utilizing this asymmetric 1,3-dipolar cycloaddition reaction, Ukaji et al. successively synthesized the lythraceae alkaloid (–)-lasubine II, isolated from the leaves of *Lagerstroemia subcostata*. The optically active 2-isoxazoline thus obtained was transformed to bicyclic lasubine II stereoselectively by sequential reduction and cyclization [74] (Scheme 27).

Successive treatment of 2-propen-1-ol with dialkylzinc, (R,R)-DIPT, a second dialkylzinc and aldoxime can afford the intermediate (**18**). Addition of halogenating agent (X–X′) leads to the oxidation of the aldoxime residue, which would coordinate to zinc as depicted in (**19**). The resulting nitrile oxide then undergoes asymmetric 1,3-dipolar cycloaddition to give the corresponding 2-isoxazoline in optically active form (Scheme 28) [75, 76].

This procedure is greatly advantageous since it does not involve the preparation of a rather unstable hydroximoyl chloride. Due to the easy availability of (*R*,*R*) and (*S*,*S*)-DIPT, this method provides a useful way to prepare both enantiomers of 2-isoxazolines which are versatile intermediates for the synthesis of optically active β-hydroxylketones or γ-aminoalcohols.

As expected, the cycloaddition of 4-*t*-butylbenzonitrile oxide with acrylamide in DMF or water afforded only the 5-substituted isoxazoline derivative

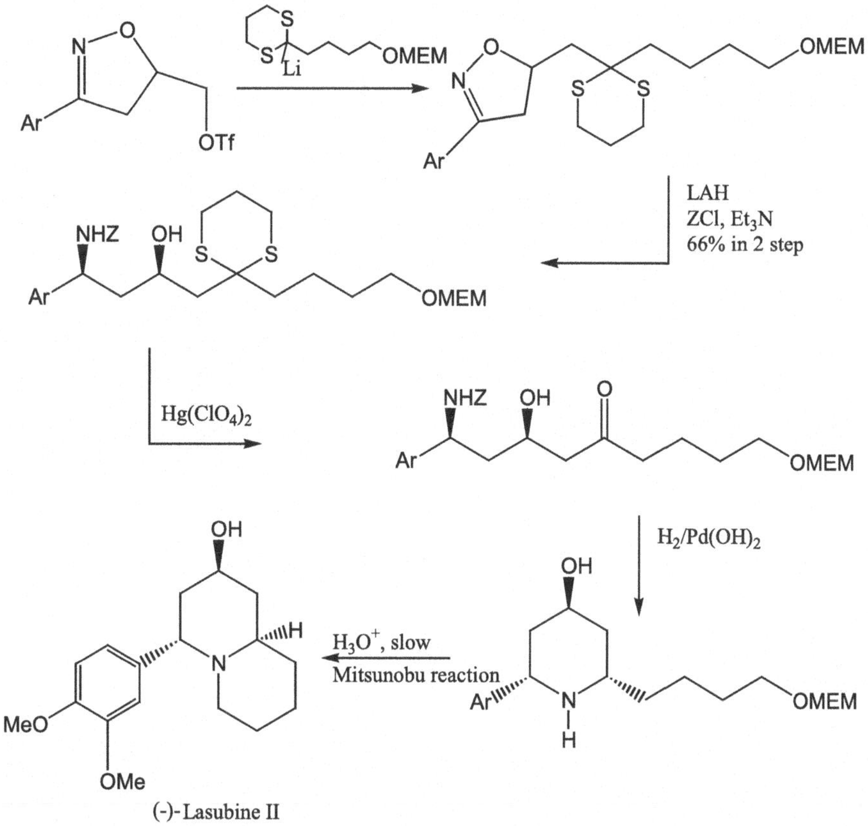

Scheme 27

(an analog of **B**). But when the same nitrile oxide reacted with acrylamide attached to cyclodextrin, it produced a 4-substituted isoxazoline **A** as the major product. It was anticipated that inclusion of the hydrophobic moiety of the dipole within the *annulus* of the modified cyclodextrin would establish a different alignment for the cycloaddition. This effect is greater in water than in DMF because the cyclodextrin inclusion complex is formed in aqueous solution (Scheme 29) [77].

Chiral cationic pentadienyl catalyst [Ru(acetone)(*R*,*R*)-BITHOP-F)Cp] [SbF_6] (**21**) was formed by the complexation of chiral ligand (**20**) with $Ru_3(CO)_{12}$, in the presence of cyclopentadiene and SbF_5. The catalyst catalyzes the (3+2) dipolar cycloaddition reaction between any nitrile oxide generated from a hydroximoyl chloride and an α,β unsaturated aldehyde to give a chiral 2-isoxazoline with 43–98% yield and 60–93% ee. The stereochemistry of the major enantiomer is S, consistent with an approach of the nitrile oxide to C_1–Si face of the enol in the *anti*-s-*trans* conformation in the catalyst site. Nitrile oxide with electron withdrawing substituents in the

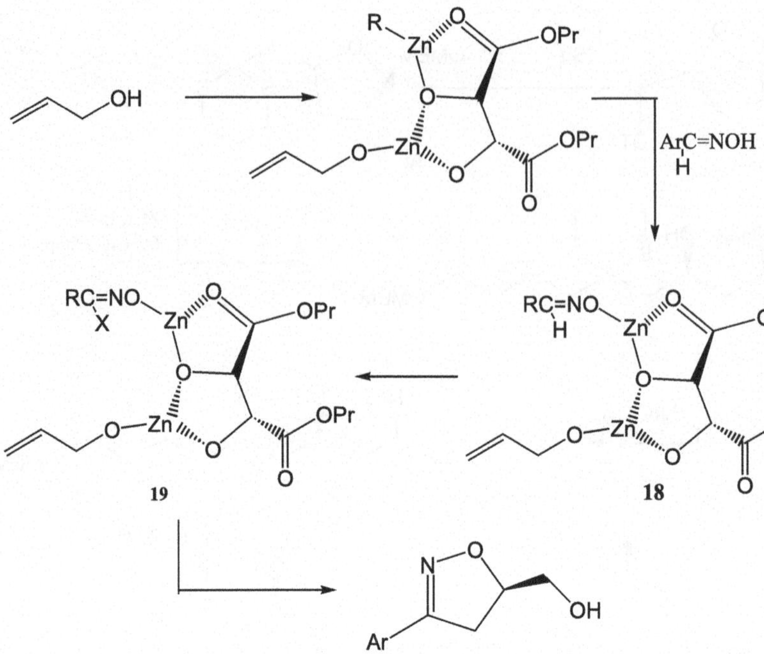

Scheme 28

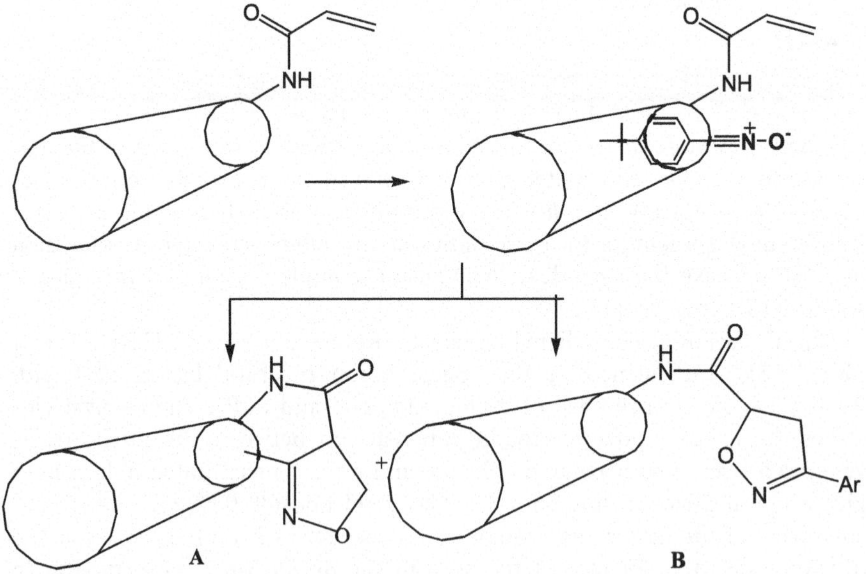

Scheme 29

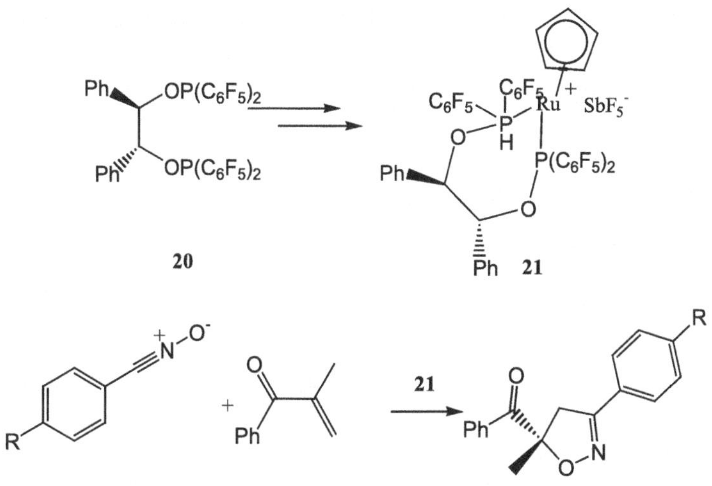

Scheme 30

4-position (CF_3, F, Cl) on the aromatic ring gave higher *ee*'s than those having electron donating (Me, *i*-Pr, OMe) groups (Scheme 30) [78].

Sibi et al. observed that a chiral Lewis acid, prepared from magnesium iodide and a chiral bisoxazoline derived from 1,2-aminoindanol, acts as an effective catalyst for the stereoselective addition of nitrile oxides to acrylamide derivatives (Scheme 31) [79].

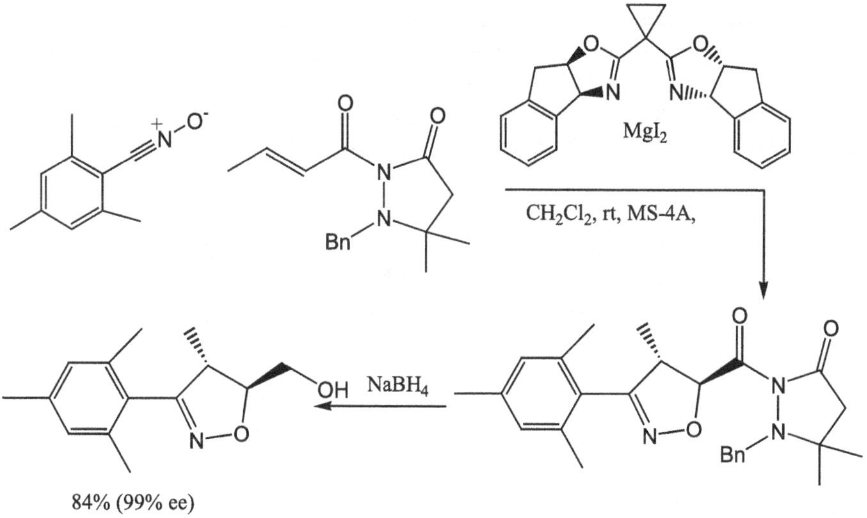

Scheme 31

Racemic α-silyl allyl alcohols can be prepared in good yields through the retro-Brook rearrangement of trimethylsilyl ether of allyl alcohol, which was generated in situ by treatment of allyl alcohol with base and TMSCl. To convert racemic alcohol (**22**) to optically active (**22**), the alcohols were oxidized to acyl silane, which were reduced with (–) DIP-Cl to give (s)-(–) **22** in 85–87% ee (Scheme 32) [80].

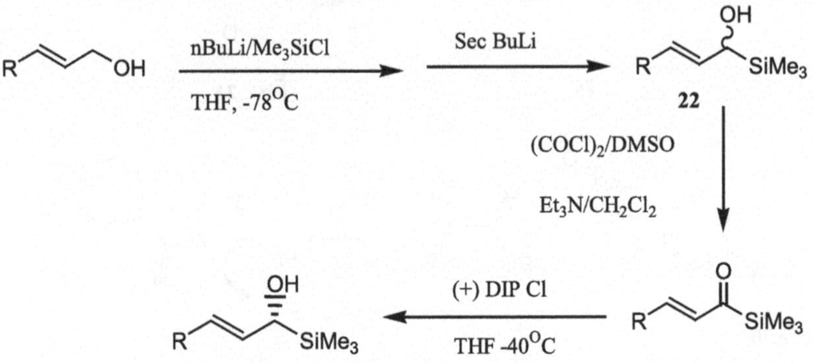

22 a. R = Pr, 75% , 82% ee
b. R =Me,24% , 85% ee

Scheme 32

The cycloaddition of aromatic as well as aliphatic nitrile oxides generated from their hydroximoyl chloride in presence of Et_3N with allyl alcohols (**22a**) gave optically active 4,5-dihydro isoxazoles in good enantiomeric excess (Eq. 3). The adduct was dextro-rotatory, hence all three chiral centers were expected to have the same configuration. This stereochemical outcome is easily understood by comparison with the reaction mechanism of the magnesium induced cycloaddition reaction (Eq. 3) [80].

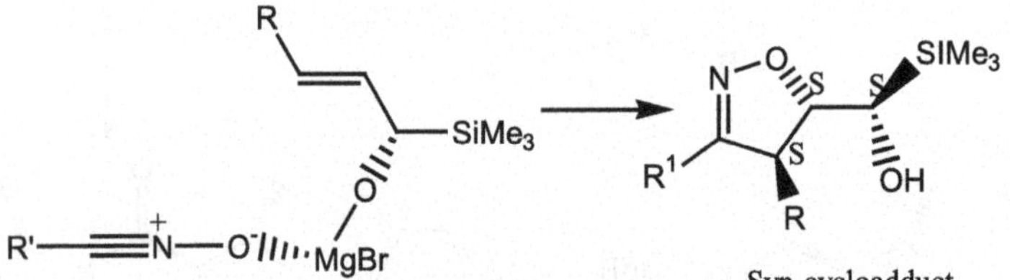 (3)

Syn-cycloadduct

When exposed to tetraphenyl ammonium fluorine, the cycloadducts were converted into the 4-substituted 5,6-dihydro-4*H*[1,2]-oxazines in good yield with loss of optical purity (Scheme 33).

Acyl silanes obtained by the oxidation of the mentioned cycloadduct, underwent stereoselective allylation reaction in the presence of a Lewis acid to lead to the elongation of C-5 side chain in a stereoselective manner (Scheme 34) (Table 3).

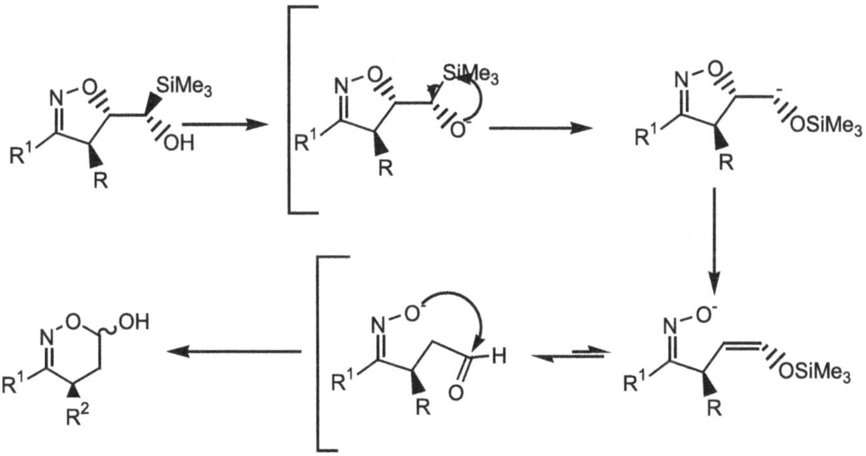

Scheme 33

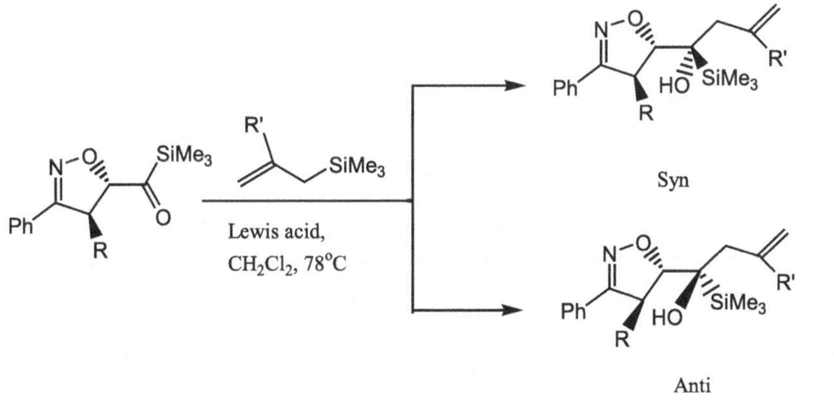

Scheme 34

Table 3 Allylation of acyl silanes

Entry	R	R′	Lewis acid	Time (h)	Yield	*Syn/anti*
1	C_3H_7–	H	$TiCl_4$	2.5	88	4/96
2	C_3H_7–	H	BF_3:OEt_2	5.0	83	7/93
3	C_3H_7–	H	$SnCl_4$	4.5	54	63/37
4	C_3H_7–	Me	$TiCl_4$	3.5	75	1/99
5	C_3H_7–	Me	BF_3:OEt_2	15.5	62	8/92
6	C_3H_7–	H	$TiCl_4$	2.0	79	20/80
7	C_3H_7–	H	BF_3:OEt_2	2.0	92	10/90

Intermolecular cycloaddition of nitrile oxides with hex-5-enopyranosides or pent-4-enofuranosides afforded spiroisoxazolines in good to excellent yields [79]. Reductive opening of the spiroisoxazolines followed by spontaneous intramolecular aldol-like condensation lead to the formation of densely functionalized 6 and 5-membered carbocycles. This transformation has afforded complimentary methodology for entry to amino cyclitols, carbosugars and ionositols (Scheme 35) [81–83].

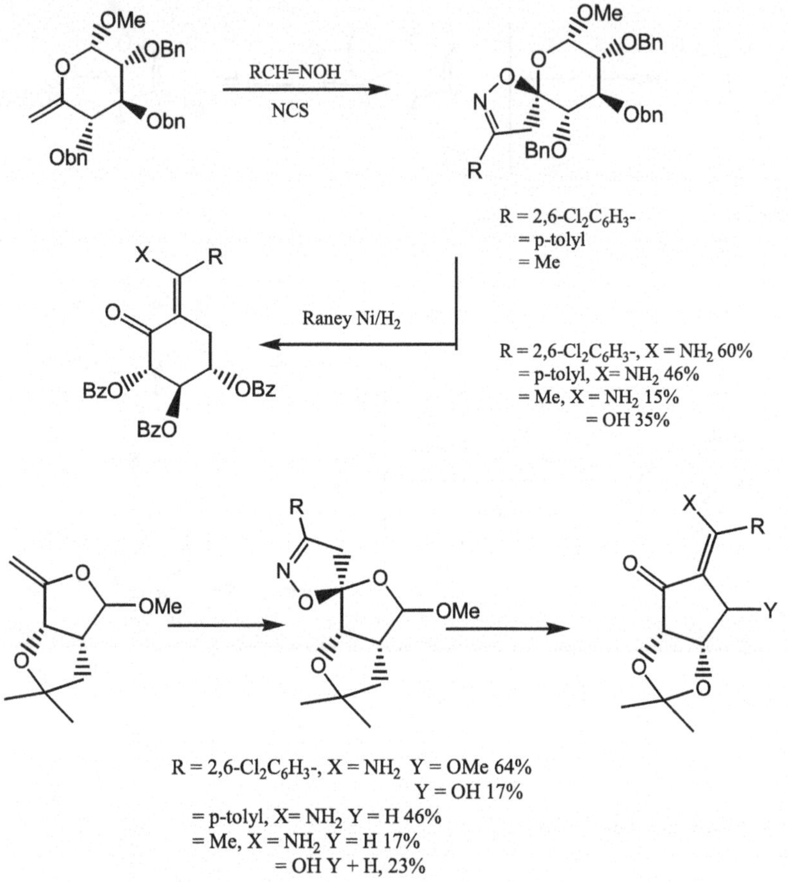

Scheme 35

Jager et al. [84] developed a conventional method for the synthesis of 2-deoxyribose via the intermolecular cycloaddition of BrCNO with an alkene derived from glyceraldehyde (Scheme 36).

Kim et al. reported the synthesis of novel crown ether type cyclophanes based on the multiple cycloaddition between bifunctional dipoles and bifunctional dipolarophiles [85]. For instance the divinyl ether derived from

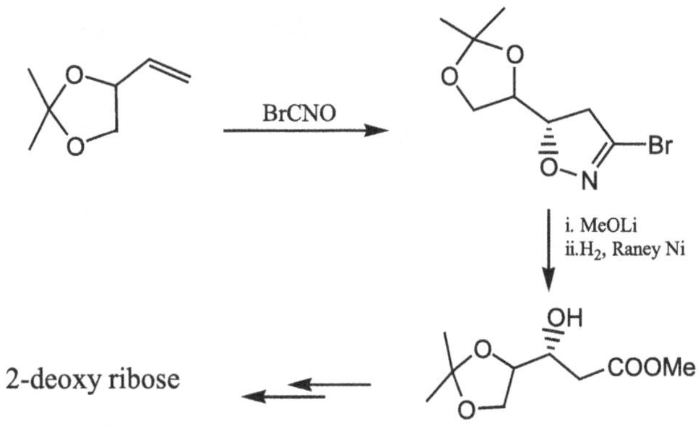

Scheme 36

triethylene glycol cycloadds with dinitrile oxide generated from the corresponding terphthalaldehyde to afford a 40-membered macrocycle in an almost quantitative yield (Scheme 37).

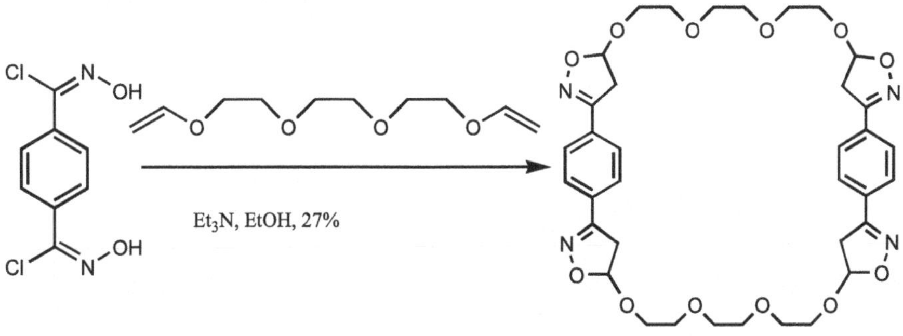

Scheme 37

Bifunctional dipoles (isophthalodinitrile oxide and 2,6-pyridine dinitrile oxide) cycloadd with diacrylates derived from triethylene glycol to afford 21-membered macrocycles (Scheme 38).

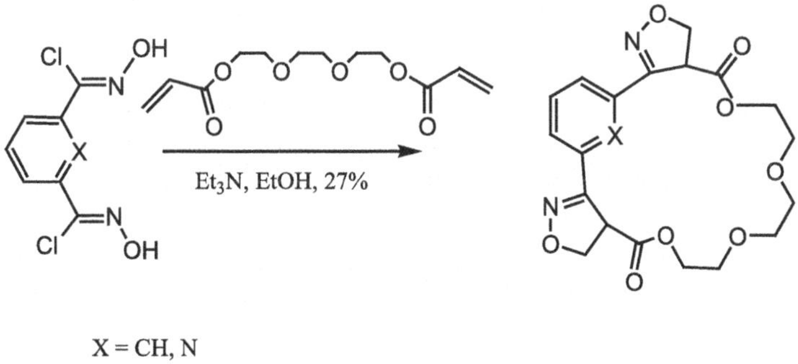

Scheme 38

In 2000, Kim et al. synthesized silicon bridged macrocycles in a two-step sequence by using quadruple cycloaddition, macrocyclization and intermolecular nitrile oxide dimerization [86]. Double cycloaddition between in situ generated isophthalodinitrile oxide and 1,3-divinyl tetramethyl disiloxane provided a [1+2] cycloadduct as the major intermediate and further cycloaddition between (**23**) and isophthalodinitrile oxide afforded the final 2+2 cycloadduct as the major product in 25% overall yield (Scheme 39). In a similar fashion, silacyclophanes were prepared by QCM methodologies between terphthalodinitrile oxide and 1,3-divinyl tetramethyl disiloxane.

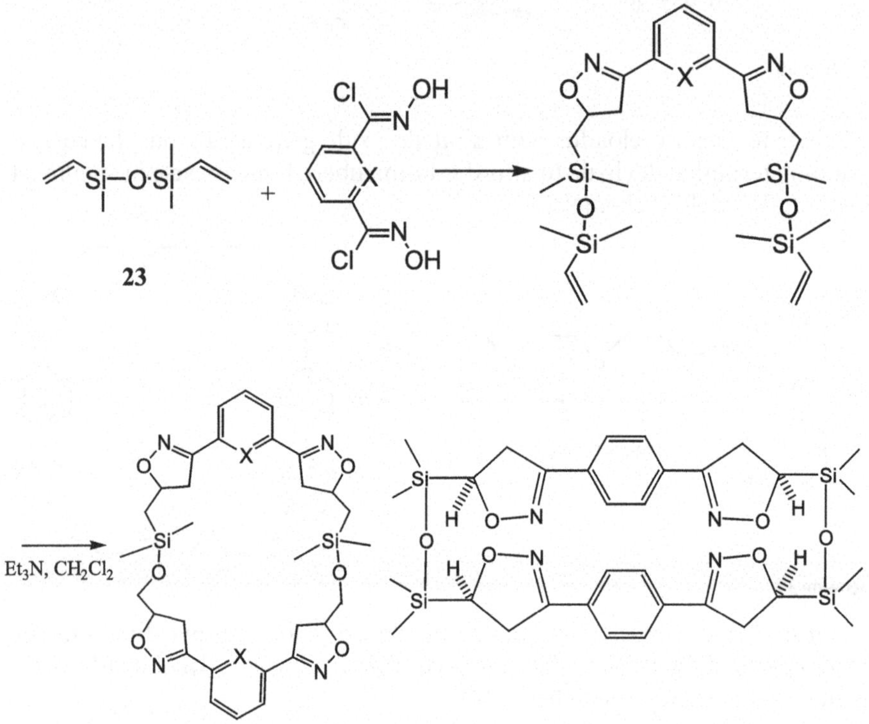

Scheme 39

Kanemasa et al. utilized alkyl magnesium bromide as a base instead of triethylamine for the generation of nitrile oxides from hydroximoyl chlorides [87]. Reaction of the resulting nitrile oxide with a chiral allyl alcohol gave (**24**) with *syn* selectivity in up to 98% yield. It was proposed that the *syn* selectivity as well as the rate acceleration observed was due to chelation via the metal salt of the two transition states leading to *syn* (**24a**) and *anti* (**24b**), the former being favored since steric repulsion between the substituent (e.g., Et) at the chiral center and the allylic proton in the *anti* transition state is to be expected (Scheme 40).

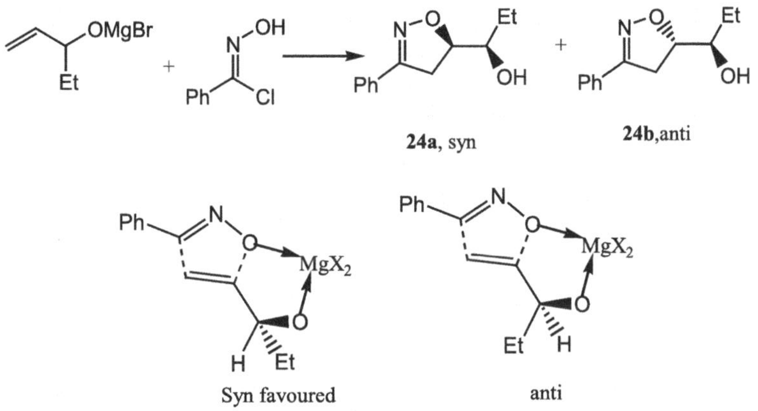

Scheme 40

Intermolecular cycloaddition of nitrile oxide generated from quinolyl hydroximoyl chloride with macrolide (**25**) produced a single compound (**26**). Ma et al. observed that the isobutylene bridged macrolide (**25**) adopts a preferred conformation, that only allows the nitrile oxide to approach from the top face of the molecule producing isoxazoline adduct (**26**) in a stereoselective fashion. Hydrolysis of the C-3 cladinose group from (**26**) followed by oxidation under Dess–Martin conditions gave the corresponding keto compound (**27**). Finally the 2′-OH protecting group was removed to give the anchor-left isomer (**27**) in good yield [88]. Molecular modeling of (**25**) suggests that the C-3 cladinose group is placed above the macrolide ring and blocks one of two pathways for the incoming nitrile oxide, leading to an anchor-left isomer (**27**) as the single product, while the (–) cladinose-free molecule allows the formation of an anchor-right isomer (**28**) (Scheme 41).

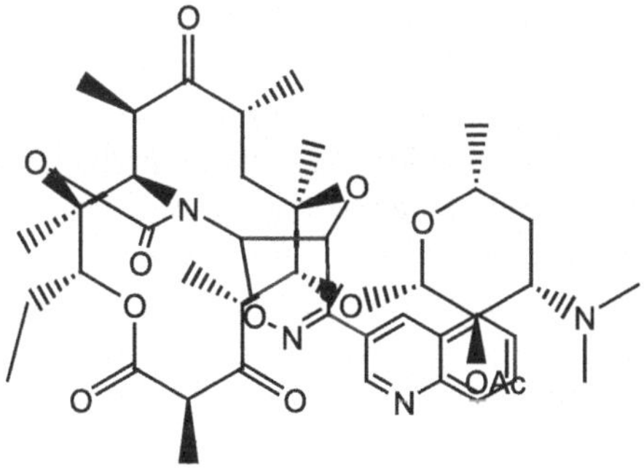

28 Anchor right

The antibacterial activity of (**27**) and (**28**) as determined by using the agar dilution method against four representative gram-positive organisms showed

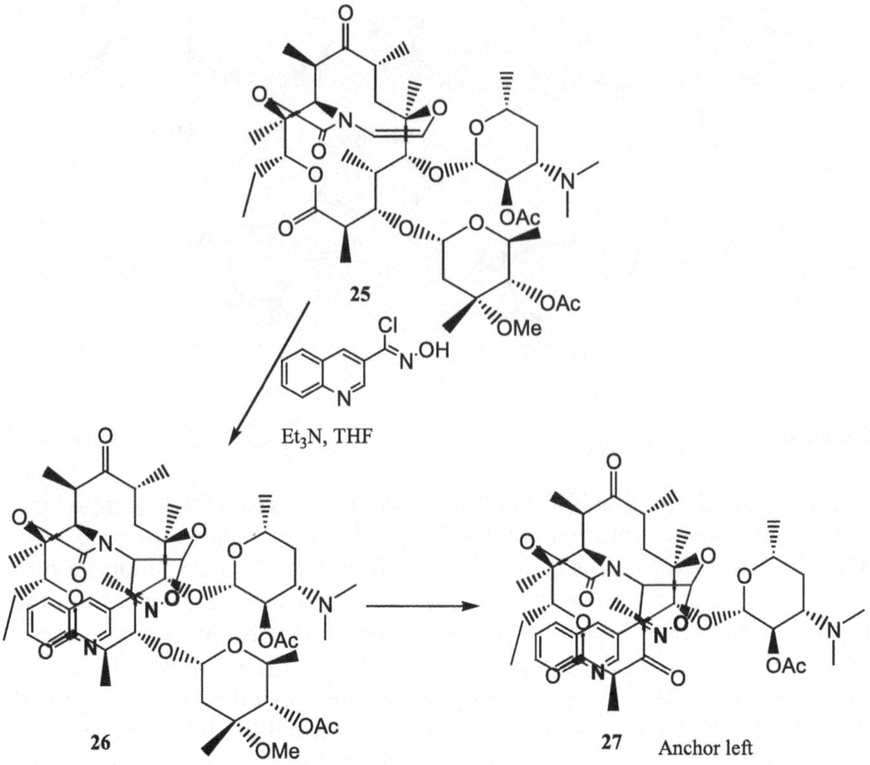

Scheme 41

that the anchor-right isomer (**28**) is about 10-fold more potent than (**27**) against the susceptible strains as well as the resultant strains with methylated binding sites. The data are consistent with a notion that the quinolyl group in compound (**28**) interacts with a secondary binding site on bacterial ribosomes, while the quinolyl group in (**27**) exerts a detrimental steric effect on the binding.

Under supercritical CO_2 ($scCO_2$) as solvent, Lee et al. demonstrated that regioselectivity of dipolar cycloaddition of mesonitrile oxide to various dipolarophiles proceed with higher stereoselectivity than in most conventional solvents. The reaction mixtures described in Table 4 are all homogeneous in $scCO_2$ under the specified condition. These results showed that cycloadditions occur with a diverse range of alkenes in $scCO_2$ to give isoxazoles and isoxazolines in high yields [89].

1.6.2 Intramolecular Cycloaddition of Nitrile Oxide (INOC)

The application of intramolecular nitrile oxide-olefin cycloaddition (INOC) reaction to the synthesis of complex natural products has recently come to be

Table 4 Cycloaddition of mesonitrile oxide to alkenes and alkynes in $scCO_2$

Dipolarophile	Time (h)	Cycloadducts	Yield (Ratio of isomers)
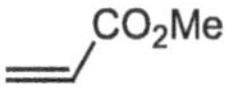	16	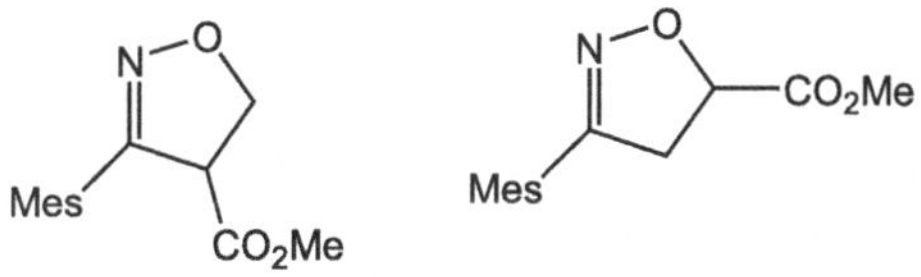	97 (8.7:1)
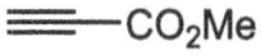	16	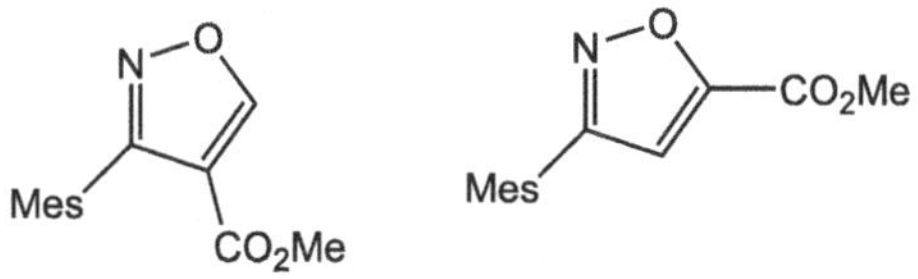	97 (2.9:1)
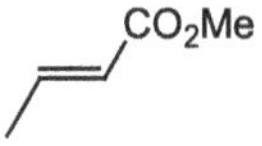	60	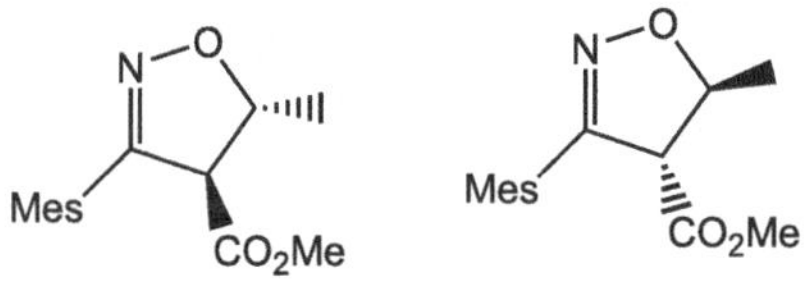	97 (4.2:1)
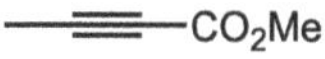	60	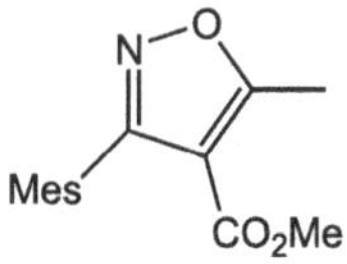	99
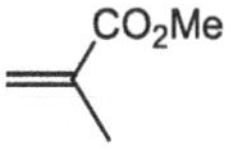	24	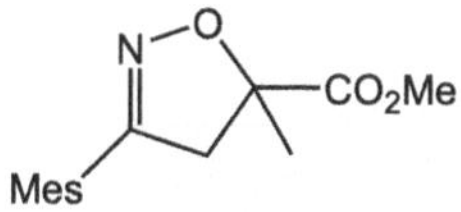	98
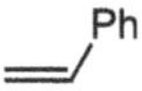	24	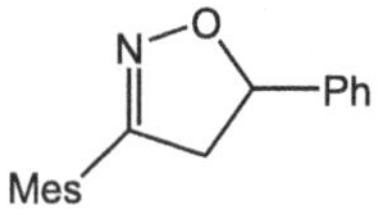	96
	24	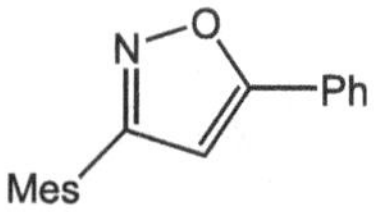	96
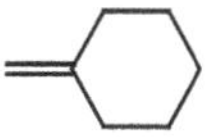	60	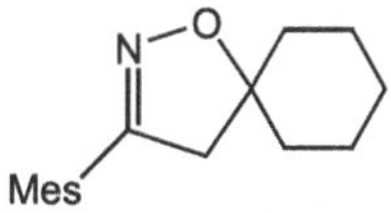	96

recognized as a powerful synthetic tool, one equally akin to the intramolecular Diels–Alder reaction in its potential scope of application. It is therefore not surprising that the INOC reaction has been extensively utilized in total synthesis. The intramolecular cycloaddition reaction generally displays exceptional regio and stereochemical control, which undoubtedly accounts for the popularity of this reaction, which can offer a powerful solution to many problems in complex natural product synthesis.

Examples of how to promote intramolecularity by means of metal chelation have been shown in Schemes 26 and 28. An alternative way involves initial covalent anchoring of the dipolarophile component to the dipole followed by the intramolecular cycloaddition.

Rai, Hassner and Dehaen [90] succeeded in getting functionalized tetrahydrofuran and tetrahydropyran derivatives by the INOC of 2-allyloxyaldoximes of type (**25**), formed by the anchoring an unsaturated alcohol to *β*-nitrostyrene in the presence of $SnCl_2$–$2H_2O$ (Scheme 42).

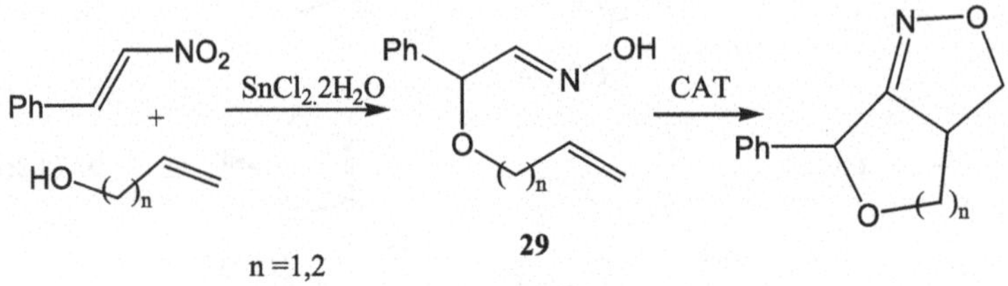

Scheme 42

Hassner et al. [91–93] showed that *O*-silyl-*α*-bromoaldoximes (**30**) represent a synthon for vinyl nitroso compounds, which can trap unsaturated alcohols, thiols or amines, leading via oximes to unsaturated nitrile oxides (**31**) tethered by an amine or ether linkage. Thus, (**30**) can be converted into an alkenylamino oxime or an alkenyloxy oxime by reaction with an unsaturated amine or alcohol, respectively, in the presence of fluoride ions (Scheme 43) [91–93]. Oxidation of such oximes (X=O) to nitrile oxides (**31**) provides a stereoselective route to functionalized tetrahydrofuran derivative

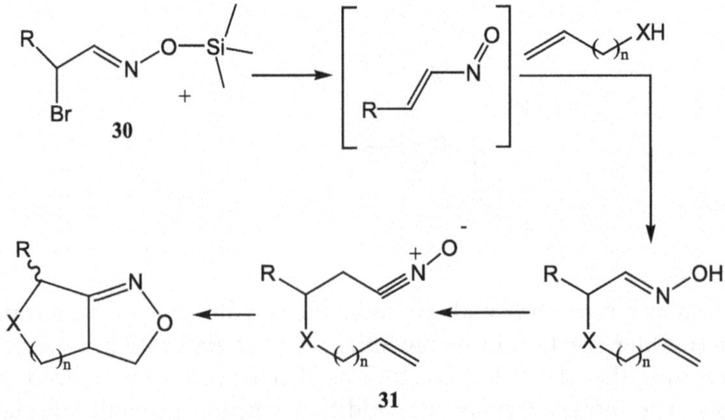

Scheme 43

via an INOC reaction [94–96]. The preferred stereoisomer in the formation of a 5-membered ring ether is *trans*, whereas in the 6-membered ring ether the *cis* isomer dominates. Similar results were obtained with unsaturated thiols (X=S) or amines (X=NR).

In a similar manner, Noguchi et al. [97] proposed a gauche–gauche interaction as a powerful factor controlling the stereochemistry in the intramolecular cycloaddition reaction of *N*-protected 3-(*N*-alkylamino)propionaldehyde (**32**) and 2-(*N*-homoalkylamino)acetaldehyde oxime (**33**). They observed high levels of stereoselectivity (76% de) in the reaction involving nitrile oxides with a substituent at the carbon atom adjacent to the tether nitrogen [97]. When the protecting group P occupies the equatorial position in the transition state of the cycloaddition of a nitrile oxide, the substituent R is expected to occupy the pseudoaxial position in order to avoid the gauche–gauche interaction between the P and R moiety. Consequently, the exclusive formation of a *syn* product results via transition state Ts-1 (Scheme 44). On the other hand, the intramolecular cycloaddition of nitrile oxide (**33**) afforded a mixture of *anti* and *syn* cycloadducts and the highest stereoselectivities (*anti:syn* = 97:3) were achieved in the reaction of (**33**), which bears a methyl group substituent (R=Me), the least sterically bulky group (Scheme 45).

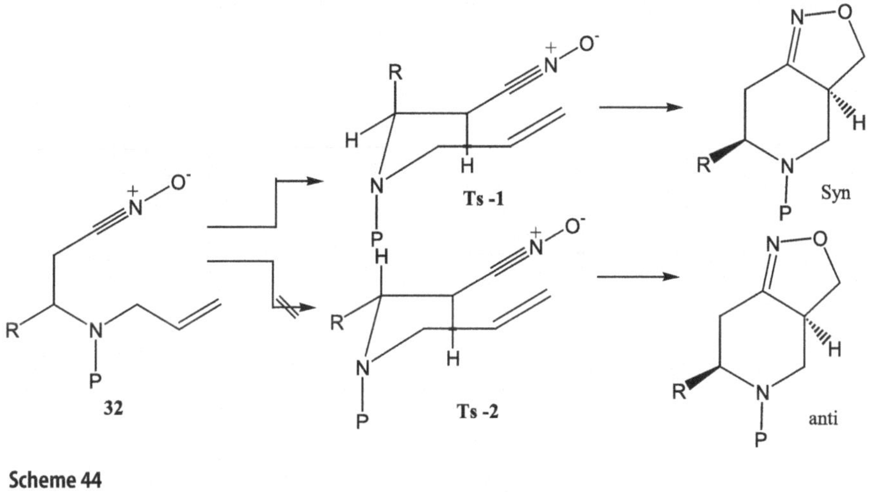

Scheme 44

Recently Scott et al. used intramolecular nitrile oxide-olefin cycloaddition reactions for constructing carbocycles, namely hexahydrobenzoisoxazoles, starting from a phenyl sulfonyl geminally substituted carbon substructure having both alkene and oxime functions [98]. The cycloaddition proceeded with up to 99% de. The authors found that geminal aryl or aryl sulfonyl moieties are considered to behave analogously to the well documented conformational bias of a tertiary butyl group. The relative stereochemistry in

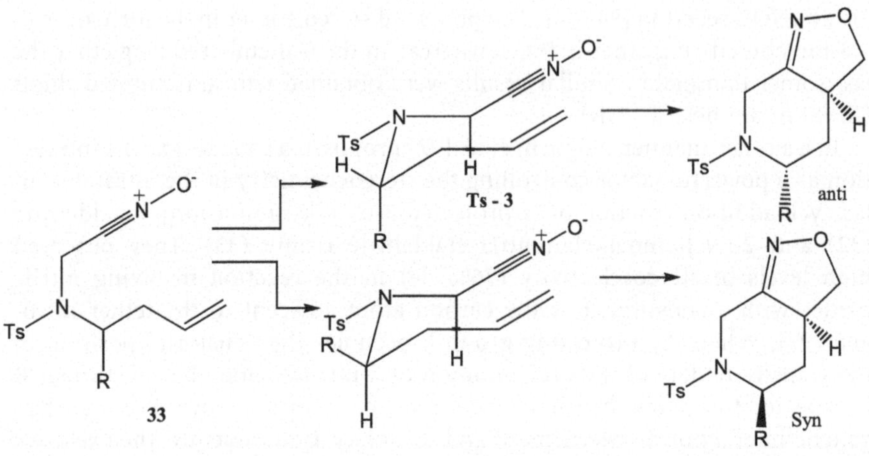

33

Scheme 45

these isoxazolines was secured by NOE studies and is consistent with the phenyl sulfonyl group occupying a pseudoequatorial orientation in preferred transition state (TS), thereby leading to (**34**) as the major diastereomeric product. Introduction of a geminal 2,5-diflurophenyl group on to the sulfone (**35**), led eventually to isoxazoline (**36**) (Scheme 46).

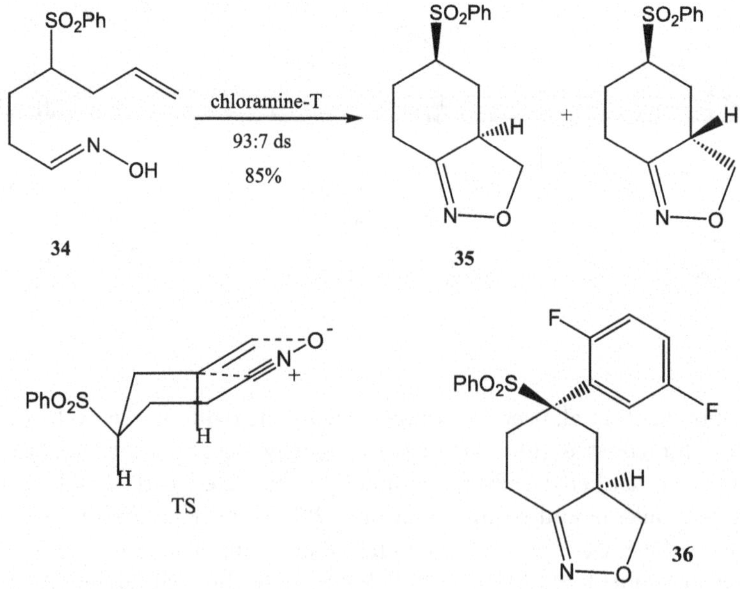

Scheme 46

Sengupta et al. [99] synthesized 3,5′-ether linked pseudooligopentose derivatives for the first time from readily available carbohydrate derivatives. These oligopentose derivatives are potentially important precursors to novel RNA analogues. Intramolecular cycloaddition of the nitrile oxide prepared from these derivatives led to the diastereoselective formation of chiral isoxazoline fused to 10 to 16-membered oxacycles (Scheme 47) [99].

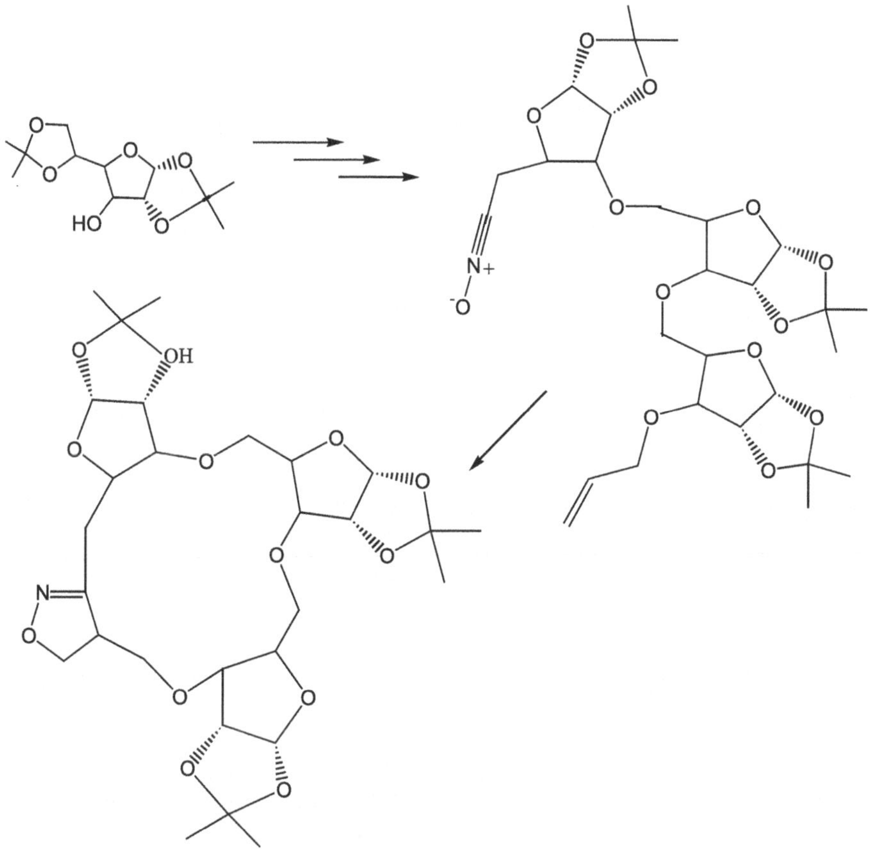

Scheme 47

Enantiomerically pure hydroxymethyl calystcginc, an alkaloid of the tropane family was synthesized starting from glucose via intramolecular 1,3-dipolar cycloaddition of nitrile oxides (Scheme 48) [100].

The enzyme secretase represents a potential target for Alzheimer disease therapeutic intervention. γ-Secretase inhibitor (**37**) represents a significant synthetic challenge. Key structural elements include a trisubstituted octahydro-1*H*-2,1-benzothiazoline-2,2-dioxide core. An important step in the

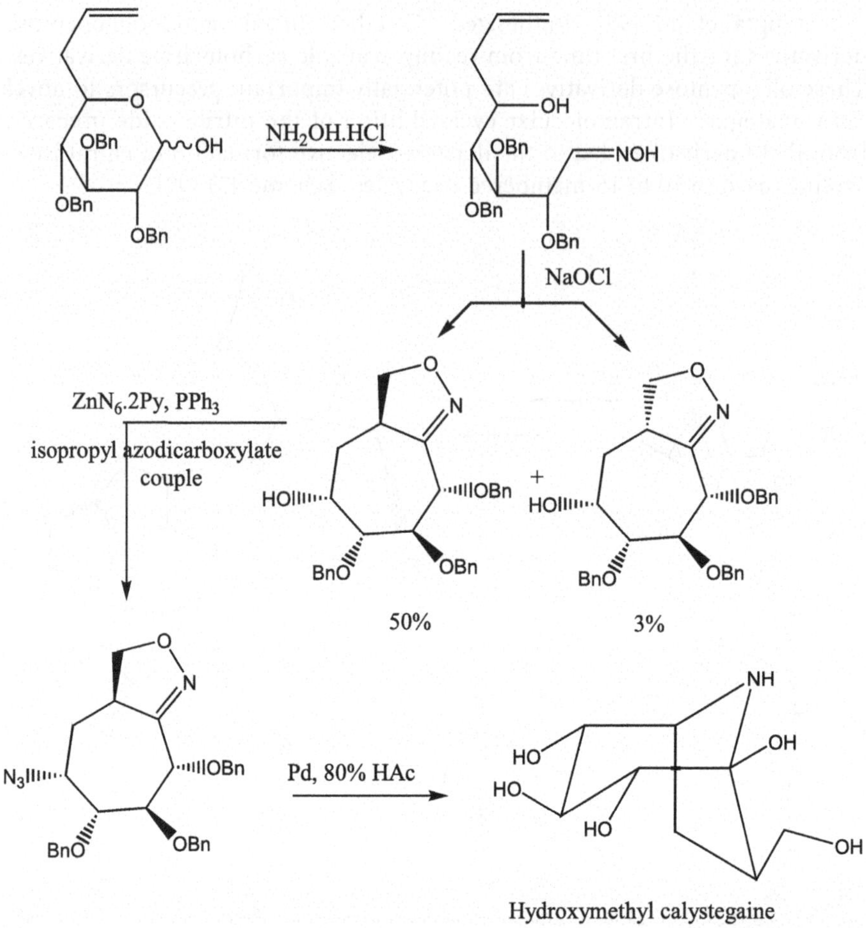

Scheme 48

synthesis of this molecules involves building carbocycles via substituent controlled intramolecular nitrile oxide-olefin cycloaddition. The reaction scheme involves ten reaction intermediates (13% overall yield) among them conversion of an aldoxime by NaOCl to a nitrile oxide followed by INOC (Scheme 49). Other aspects of the route include a highly stereoselective reduction of an isoxazole to form a *cis*-γ-aminoalcohol, an efficient chemical resolution, a dianion cyclization to construct a sultam ring and the α-alkylation of a sultam with excellent diastereoselectivity [101].

Intramolecular nitrile oxide-olefin cycloaddition of sugar derivatives with unmasked hydroxy groups employing chloramine-T in the presence of silica gel proceeded smoothly to afford 5 or 6-membered carbocycle in excellent yields. This methodology, based on the finding by Rai and Hassner [28], obvi-

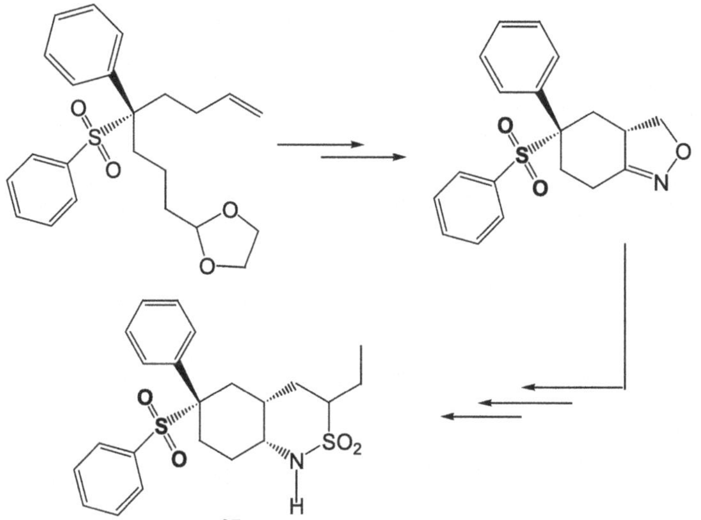

37

Scheme 49

ates protection/deprotection steps and makes the synthetic route simpler and more efficient (Eq. 4) [102].

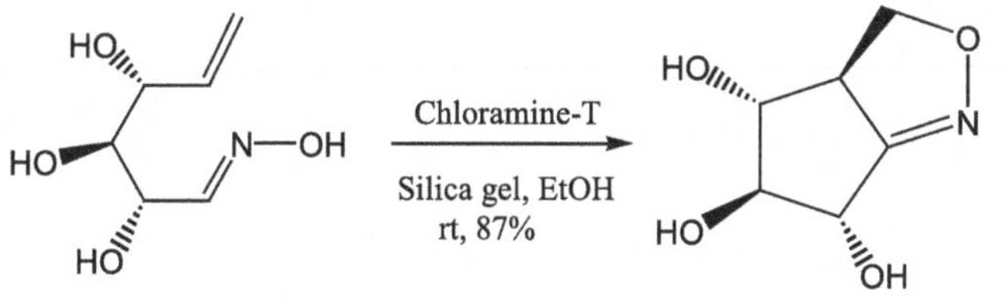

(4)

A nitrile oxide generated from a sugar-derived aldoxime (**38**) underwent intramolecular nitrile oxide cycloaddition reaction to chiral pyranoisoxazoline derivative (**39**) (Scheme 50). Reductive cleavage of isoxazoline (**39**) with LAH followed by acetylation provided the polysubstituted pyran derivative (**40**) [103].

Gallos and Koumbis reviewed the use of nitrile oxides derived from aldoximes in the synthesis of carbohydrate mimics via 1,3-dipolar cycloaddition reactions [104]. While formation of 7-membered ring ethers was not successful in simple systems as in Scheme 42, it was possible to obtain oxepine derivatives by making the system more rigid and attaching the nitrile oxide and olefin groups to a 5-membered ring. Thus, Shing et al. constructed fused oxepine derivative (**42**) via intramolecular cycloaddition of an alkenyl

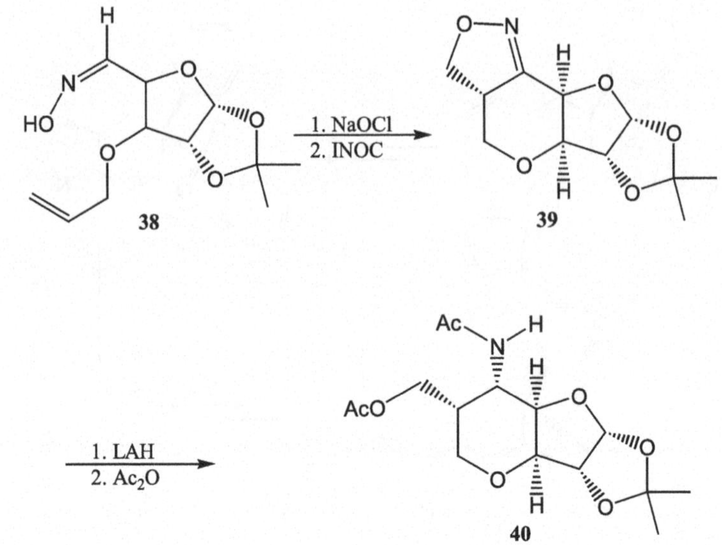

38 39 40

Scheme 50

nitrile oxide prepared from glucose-derived oxime (**41**) in 65% yield as the sole isomer but with a de of only 13% (Eq. 5) [102].

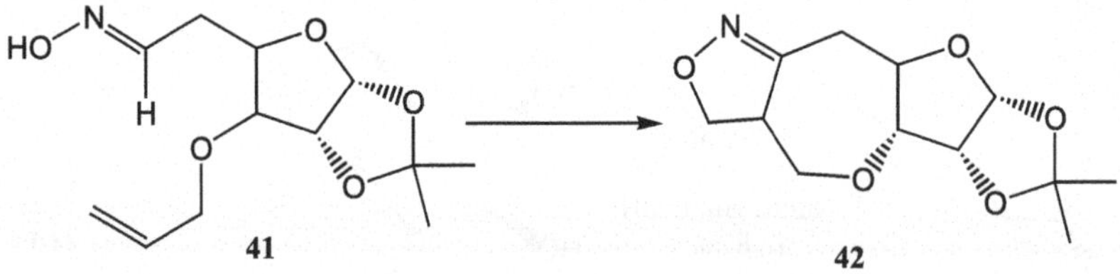

41 **42** (5)

Alkenyl oxime derived from mannitol underwent INOC reaction upon treatment with NaOCl at room temperature to afford cycloadduct (**43**) in a near quantitative yield (Eq. 6) [105]. This product was used for the synthesis of vitamin D_3. Based on theoretical calculation, it is proposed that the cycloaddition proceeds via a chair-like transition state.

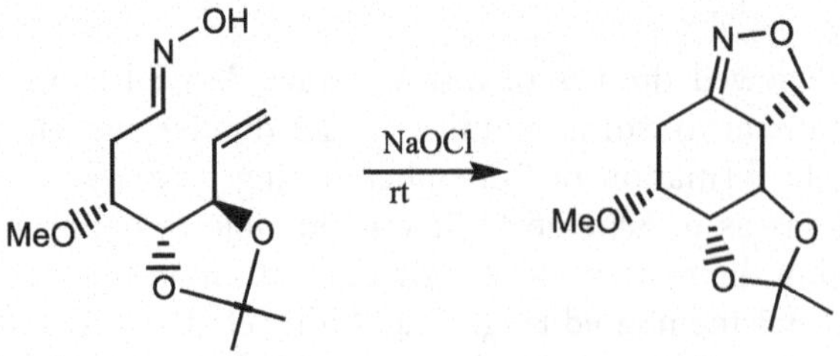

43 (6)

The key step for the synthesis of the polyol segment of the heptaine macrolide antibiotic *amphotericin* was the 1,3-dipolar cycloaddition of nitrile oxide, generated from aldoxime, to the isoxazoline (**44**) (Scheme 51) [106].

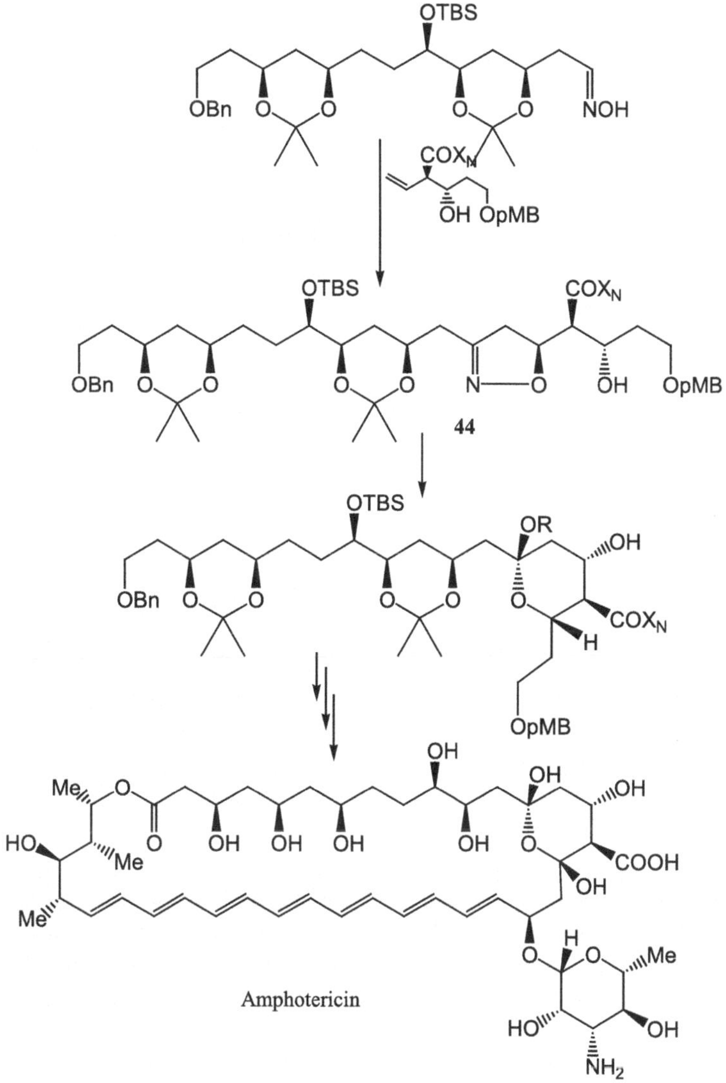

Scheme 51

The synthesis of chiral 10 to 12-membered nitrogen and oxygen heterocycles fused to isoxazoline rings was achieved by intramolecular cycloaddition of nitrile oxides derived from carbohydrate based oximes (Scheme 52) [107].

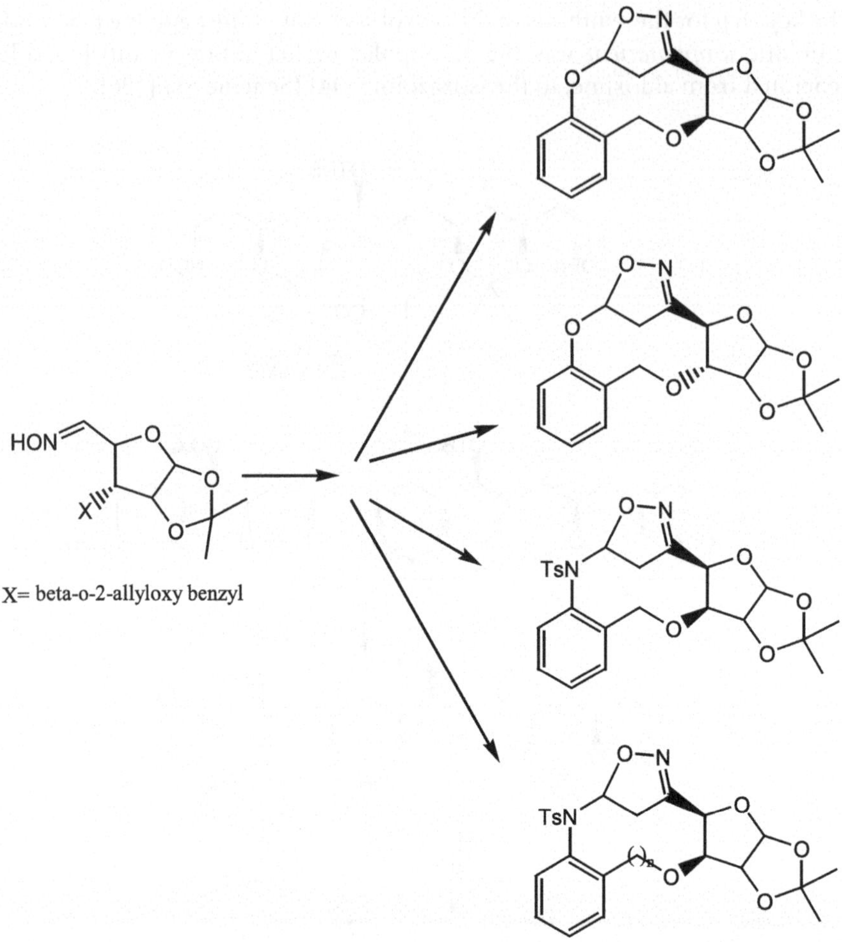

Scheme 52

Alternatively, enantiopure pyranoisoxazole and oxepinoisoxazole analogous can be prepared from carbohydrate derivatives via the corresponding nitrile oxide [108, 109]. This has allowed entry to benzopyrans, benzoxepines and functionalized tetrahydrofurans [110]. A new approach to the synthesis of the aglycon portion (**45**) of calicheamicin [111, 112], an anticancer antibiotic possessing phenomenal anticancer properties, was reported by Nicolaou et al. [112] and is based on an intramolecular alkenyl nitrile oxide dipolar cycloaddition reaction which leads directly to the incorporation of the full functionality of the aglycon. The key step for the synthesis of hexahydronaphthalene portion of the hypocholesterolemic agent, compactin (**46**) [113] involves the INOC reaction of the corresponding cycloakenyl oximes.

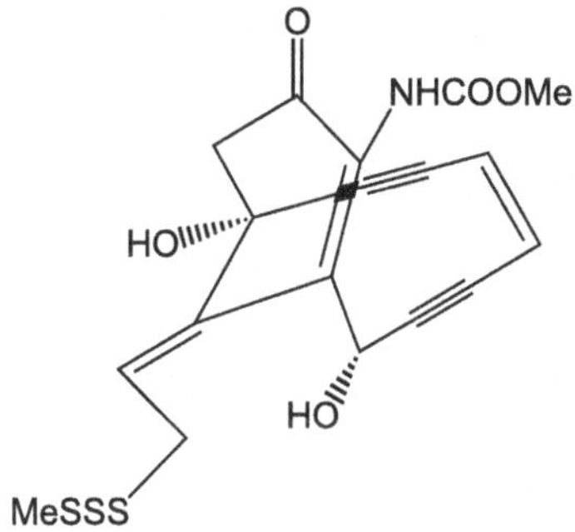

45

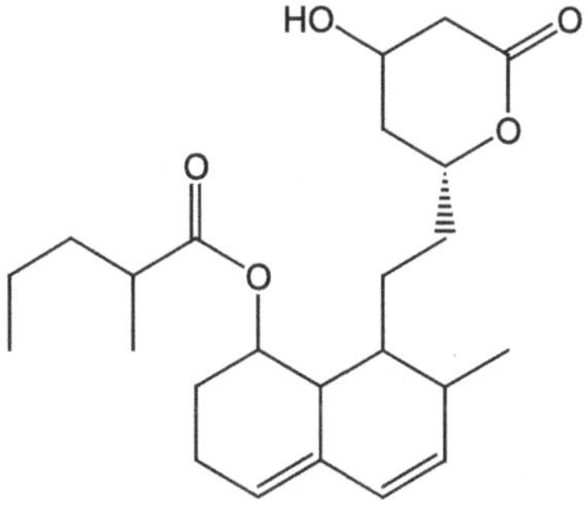

46

A related strategy was utilized for the stereoselective synthesis of (+) testosterone (**50**) wherein the A/B ring system was constructed via INOC reaction of (**47**) to isoxazoline (**49**) (Scheme 53) [114]. The cycloaddition was assumed to be taking place via chair-like TS-**48** providing isomerically pure isoxazoline (**49**) in 87% yield.

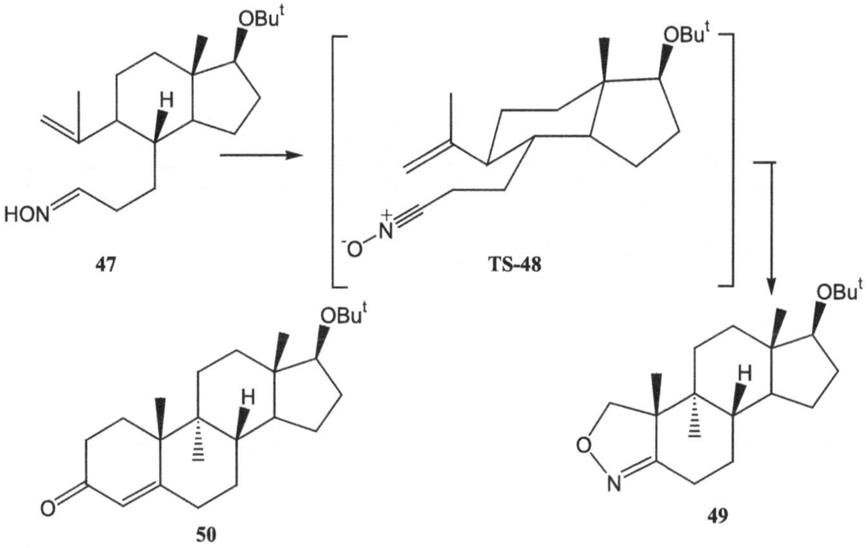

Scheme 53

Hassner et al. [115–117] utilized intramolecular nitrile oxide cycloaddition reaction as a key step for the stereoselective synthesis of ptilocaulin, a potent antileukemic and antimicrobial agent isolated from marine sponges (Scheme 54).

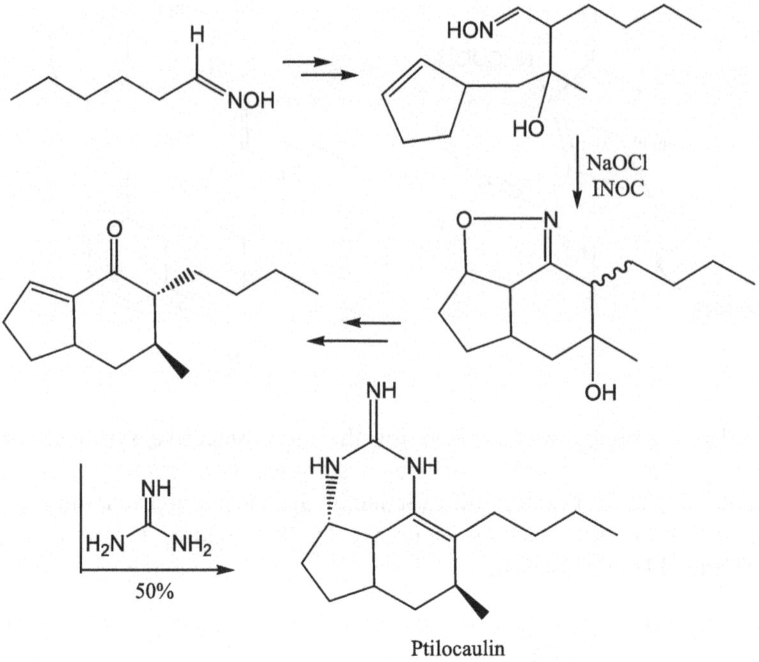

Scheme 54

Prompted by the biological importance of the 2-isoxazoline moiety coupled with the chromane ring system, Das et al. was able to synthesize 3a,4-dihydro-3*H*-chromeno[4,3-c]isoxazoles via an intramolecular nitrile oxide-olefin cycloaddition reaction employing ceric ammonium nitrate as dehydrogenating agent (Eq. 7) [118].

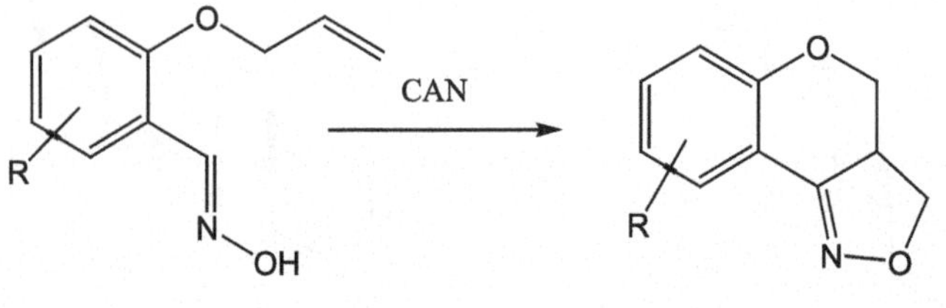

(7)

Intramolecular cycloaddition of a nitrile oxide generated from the propargyl oligoether (**51**) yielded heterocyclic crown ether (**52**) in moderate yield (Scheme 55) [119].

Mateos et al. [120] reported the stereoselective synthesis of a 12-acetoxy azadiradione analog (**53**) related to limonoid azadiradione in 12 steps starting from tricyclic derivatives (Scheme 56). The key steps for this involves intramolecular 1,3-dipolar cycloaddition of a nitrile oxide-olefin and a Stille coupling reaction of a vinyl iodide with stannylfuran [120].

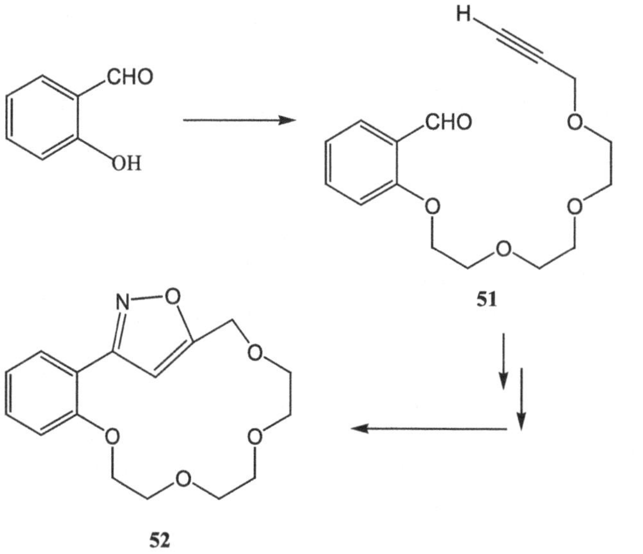

Scheme 55

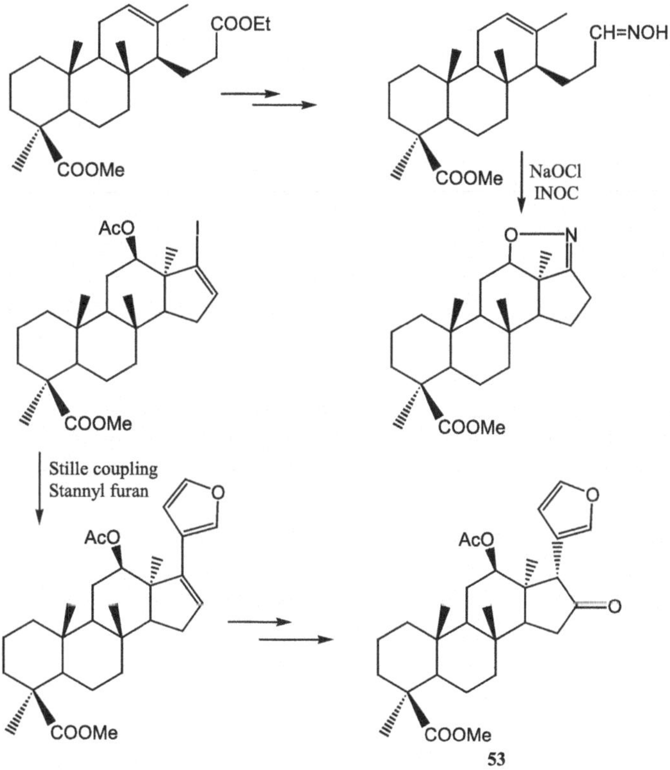

Scheme 56

Enantiomerically pure annulated sulfones having fused isoxazolines were synthesized by intramolecular cycloaddition of a chiral nitrile oxide generated in situ from the corresponding oxime by oxidation using NaOCl (Eq. 8) [121].

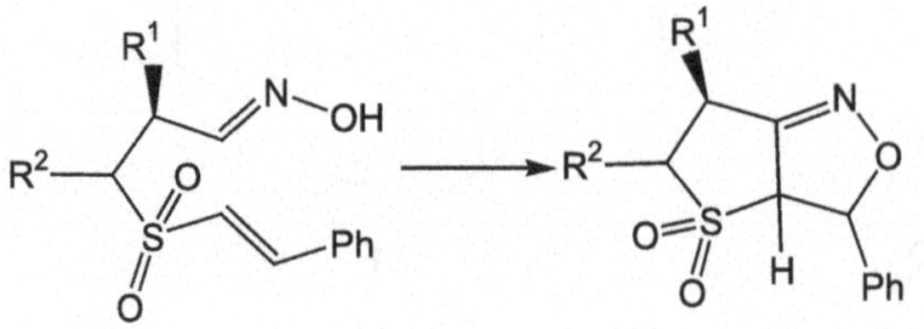

R^1 = Br, R^2 = Me 80% 100%
R^1 = Me, R^2 = Br 99% 1% (8)

A key step for the synthesis of antitumor agent pacitaxel involves the construction of the C-ring of the molecule. This was easily achieved by Takahashi et al. employing intramolecular cycloaddition of in situ generated alkenyl nitrile oxide from the corresponding oxime (Eq. 9) [122].

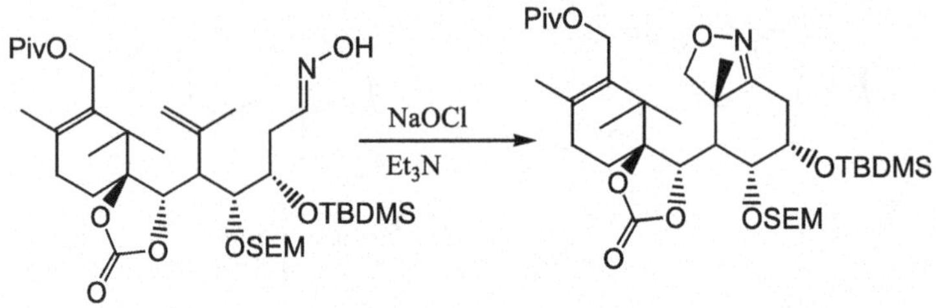

(9)

Structurally unique polypeptide macrolide, epothilone A and B isolated from cultures of *Sorangium cellulosum* has the ability to induce tubulin polymerization and associated cellular effects. Their unique structure and important biological activity conspired to make the epothilones attractive targets for total synthesis. Carreira et al. developed the intermolecular 1,3-dipolar cycloaddition of nitrile oxides as a strategy to achieve a concise, convergent and fully stereocontrolled approach for the synthesis of the above polypeptide macrolide (Scheme 57) [123]. The described route is part of an extensive study of nitrile oxide-olefin cycloaddition as a surrogate for the aldol addition reaction and has led to the realization of a higher convergent synthesis based on Kanemasa hydroxy-directed nitrile oxide cycloaddition.

The induction of enantioselectivity in 1,3-dipolar cycloaddition of nitrile oxide with alkenes was also achieved by the use of baker's yeast. Though enantioselectivity obtained was low, it can be increased up to 64% ee by the addition of cyclodextrins (see Scheme 29). In the presence of cyclodextrin, 4-vinyl pyridine first forms an inclusion compound. To this complex in water, the mixture of a stable nitrile oxide and baker's yeast in a buffer solution was

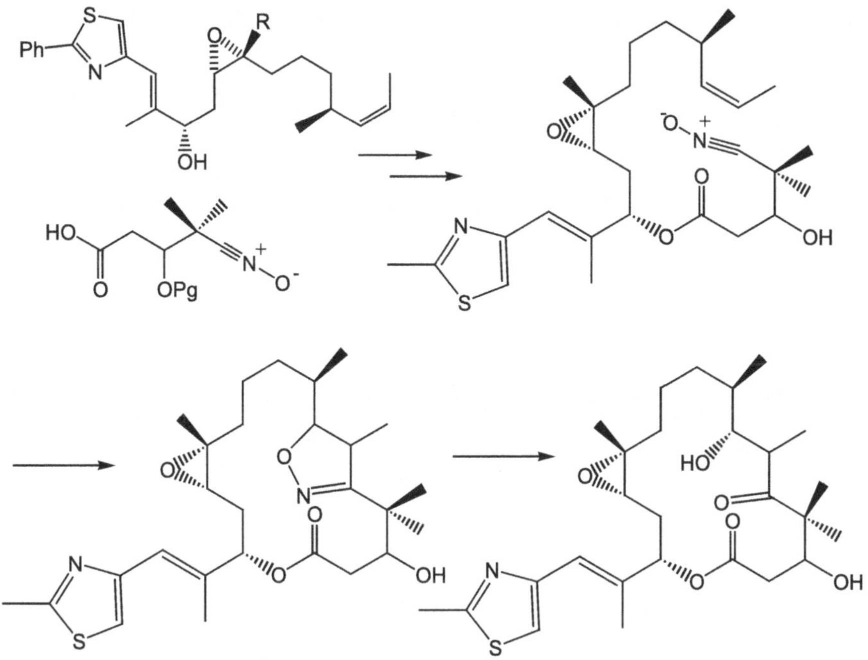

Scheme 57

added. After incubation at 37 °C for 20 h, the product is obtained in 85% yield with an optically purity of 64% ee (Eq. 10) [124].

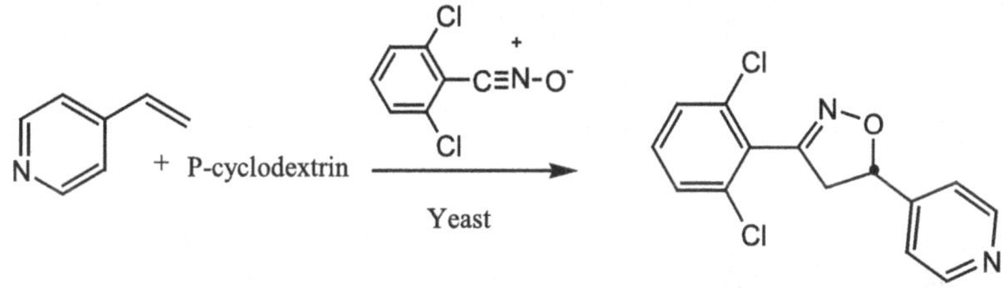

(10)

A key step for the stereoselective synthesis of Illudin C, an sesquiterpenoid possessing anticancer activity, involves the coupling of dianion (**54**) with cyclopropyl ketone (**55**) followed by intramolecular nitrile oxide cycloaddition reactions (Scheme 58) [125]. Funk et al. observed a *cis* stereochemical relationship between the C-4 methyl substituent and the C-2 hydrogen atom in the resultant tetracycle. This diastereoselectivity is a consequence of preferential cycloaddition of the nitrile oxide through a conformer wherein the smaller C(4)-hydroxyl substituent, vis-à-vis the C(4) methyl substituent occupies an equatorial position bisecting the cyclopropane ring system.

The β-glucosidase inhibitors cyclophellitol (**59**, X=O), nagstatin and gualamycin, which are microbial metabolites and their analogs have been

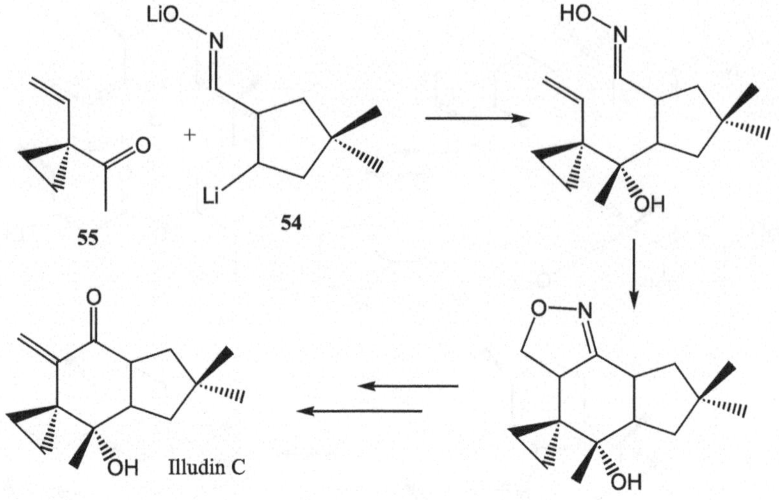

Scheme 58

synthesized from carbohydrate derivatives via 1,3-dipolar cycloaddition reactions. The synthetic scheme involves Swern oxidation of (**56**), which was derived from L-glucose and afforded the unstable aldehyde. Wittig alkenation of this aldehyde with methylene triphenylphosphorane followed by hydro-

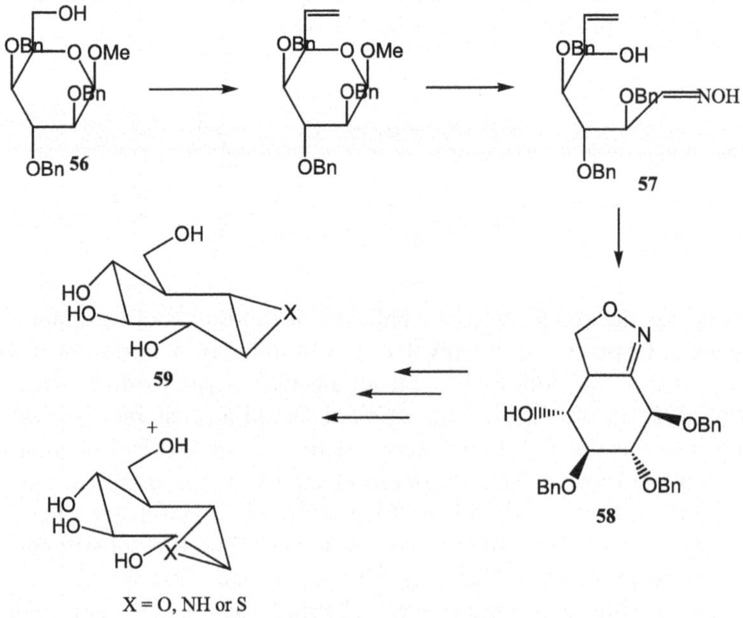

Scheme 59

lysis afforded alkenyl aldehyde. Intramolecular cycloaddition of the nitrile oxide derived from aldoxime (**57**) was realized by using NaOCl to afford the isoxazoline (**58**) as a single product in 70% yield (Scheme 59) [126, 127].

The intramolecular cycloaddition reaction was used to prepare compound (**60**) [128], the optically active intermediate in the synthesis of zoaptanol (**61**), one of the novel diterpenoids containing an oxepane ring possessing contraceptive activity (Scheme 60) [129].

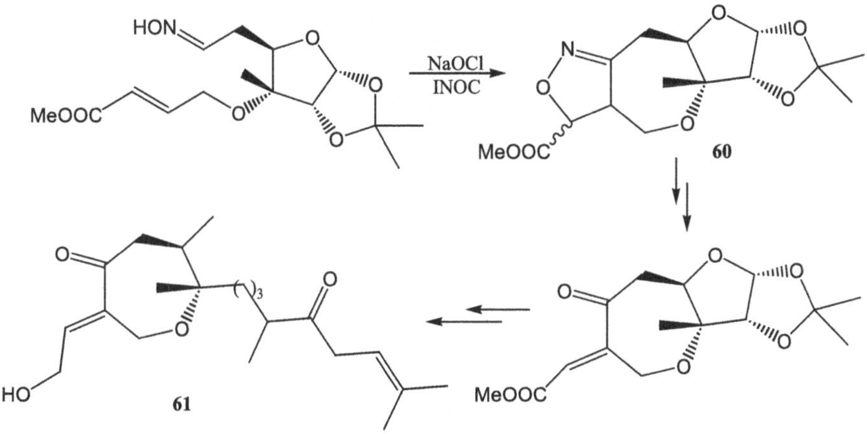

Scheme 60

The reaction of α-bromoaldoxime with unsaturated alcohols has been extended to the heterocyclic system in which furans, thiophenes, pyrroles or indoles can act as the unsaturated component of the INOC reaction [130, 131]. Equation 11 shows some examples of intramolecular cycloaddition of nitrile oxides derived from furanyl and thiophenyl oximes which led to the formation of unsaturated tricyclic isoxazolines in high yield. In this case the heterocycles ring acts as the dipolarophile with one of the double bonds adding to the nitrile oxide [130, 131].

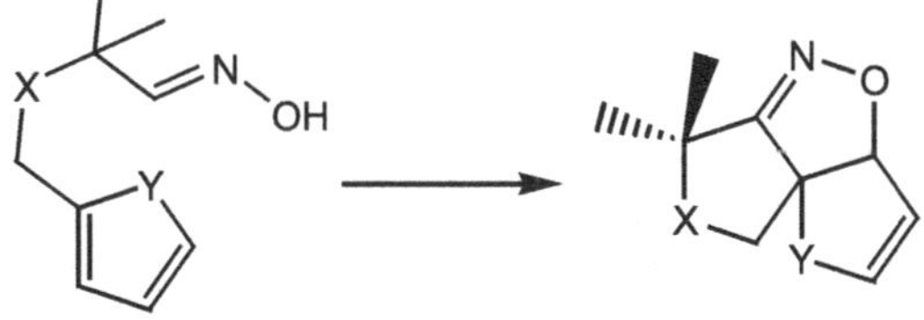

(11)

a) X=Y=O
b) X=O, Y=S
c) X=S, Y=O
d. X=N, Y=O

Hassner et al. observed that alkenyl oxime (**62**) (R_1 = H), obtained from an α-bromoaldoxime, undergoes INOC reaction on treatment with NaOCl to afford tricyclic products as mixtures of diastereomers. The major product, explained on conformational grounds, was (**64**) the stereoisomer possessing CH_2O and Me *trans*, while the R and Me groups were in a *cis* configuration (in both **63** and **64**). Oxime (**62**) ($R{=}R^1{=}Me$) failed to form cycloaddition products since in this case one of the methyl group would necessarily interfere in the transition state for cyclization (Scheme 61) [132, 133].

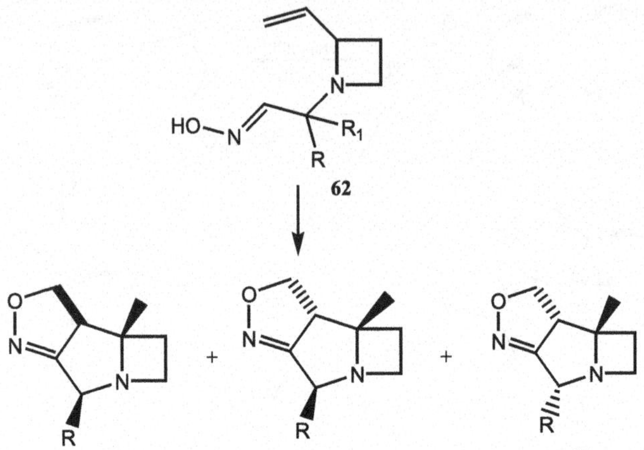

Scheme 61

Nitrile oxides are reported to react with hydrazines to give hydrazidoximes, while with alkanone hydrazones they afford the cycloaddition product, 4-amino-3-aryl-5,5-dialkyl-4,5-dihydro-1,2,4-oxadiazoles (Scheme 62) [134].

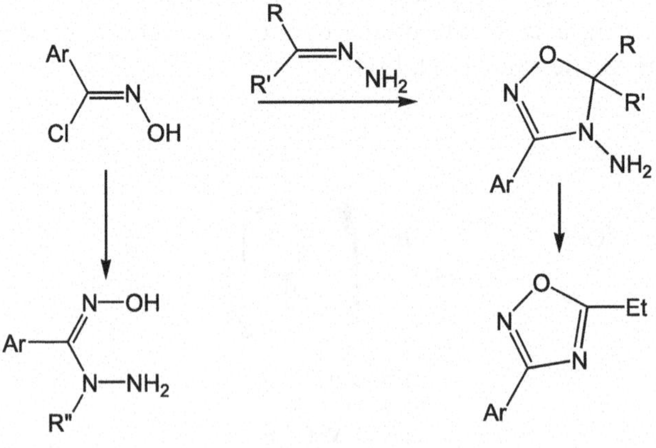

Scheme 62

1.6.3
Intramolecular Oxime-Olefin Cycloaddition Reactions (IOOC)

Grigg et al. proposed a new concept concerning the isomerization of an oxime to a nitrone through the thermal 1,2-hydrogen shift [135, 136]. Though nitrones are relatively stable species that can undergo cycloadditions with olefins, no examples of the isolation of NH-nitrones (e.g., structure **66** in Scheme 63) have been reported. Since tautomerization from an oxime to an NH-nitrone is a thermodynamically unfavorable process, NH-nitrone-olefin cycloaddition is expected to proceed only under high temperature conditions. Yet, trapping of such fleeting intermediates intramolecularly should be more favorable, and indeed intramolecular oxime-olefin cycloaddition (IOOC) to give fused isoxazolidine derivatives can be achieved (see Scheme 63). Some of the seminal studies on the IOOC reaction were reported by the Hassner group [137–140]. Modest to excellent levels of diastereocontrol were observed in the cycloaddition reaction depending on the nature of the oxime precursor. IOOC reactions differ from INOC reactions in as much as they lead to the completely saturated isoxazolidines. In both cases oximes are the starting materials. While nitrone-olefin cycloaddition leads to *N*-substituted isoxazolidines, the IOOC reaction (via a NH-nitrone) affords the *N*-unsubstituted analogs and can be more attractive because an oxime functionality is often more readily available and stable than a nitrone.

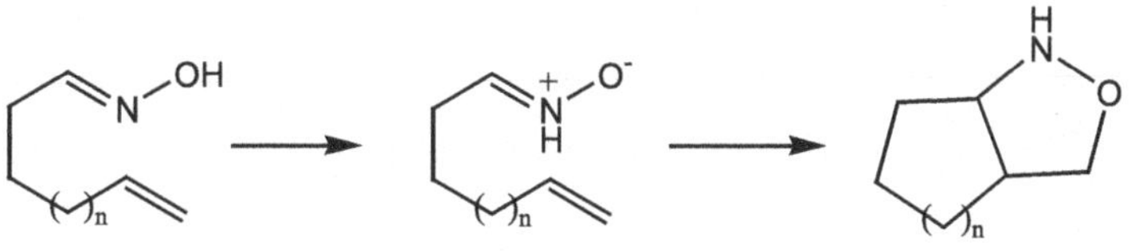

66

Scheme 63

β-Alkenyloxy aldoxime (**29**) formed by the reduction of *β*-nitrostyrene with $SnCl_2$–$2H_2O$ in the presence of unsaturated alcohols undergo thermally induced intramolecular oxime-olefin cycloaddition at 110–120 °C to bicyclic isoxazolidines (**67**) in 70–80% yield. This ring closure proceeded stereospecifically to generate three adjacent stereogenic centers. LAH reduction of (**67**) resulted in isolation of stereospecifically functionalized tetrahydrofuran derivatives in 75% yield (Scheme 64) [94–96].

β-*N*-Allylamino aldoximes prepared from silylated *β*-bromoaldoxime and allylamines undergo smooth intramolecular cycloaddition to the pyrrolidinoisoxazolidine (**68**) in 65–100% yield, simply by heating at 80–100 °C or often even upon standing for a long period of time at room temperature (Scheme 65) [141]. LAH reduction of (**68a**) and (**68b**) led to the isolation

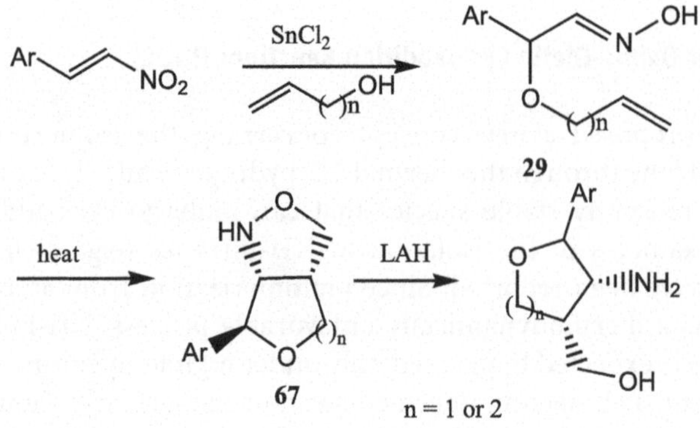

Scheme 64

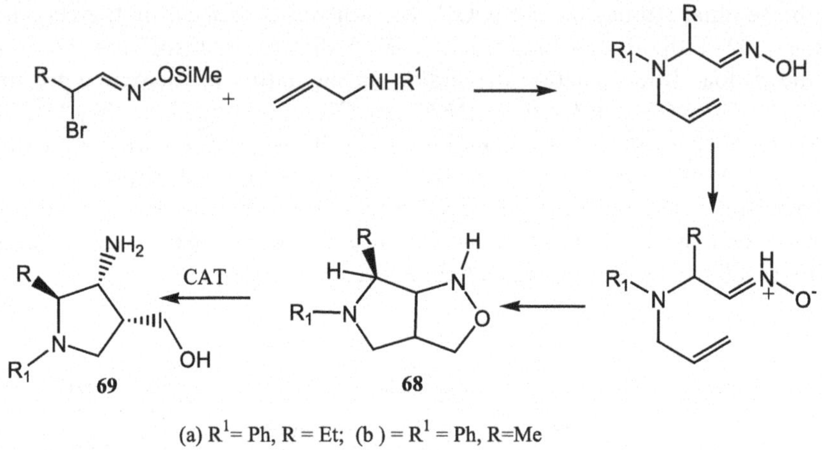

Scheme 65

of stereospecifically functionalized pyrrolidine derivatives (**69a**) and (**69b**) in 75% and 82% yield, respectively.

Furthermore, the scope of these oxime-olefin cycloadditions has been extended to ketoximes. Heating of *N*-alkylamine ketoxime at 110 °C for 8 h led to cycloaddition with formation of fused pyrrolidine (**70**) in 88% yield. Here only one stereoisomer was formed and LAH reduction led stereospecifically to the amino alcohol shown in Scheme 66.

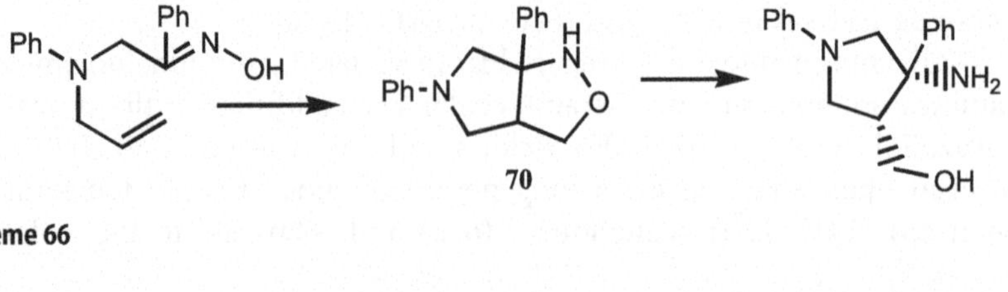

Scheme 66

In order to evaluate the stereoselectivity and the conformational requirements of the IOOC cycloaddition, Hassner et al. attempted to synthesize pyrrolidine-fused isoxazolines by heating variously substituted allylamine aldoximes (Eq. 12) [142]. The presence of terminal (γ) methyl substituents on the allylamine enhanced the reaction rate while methyl substitution on the β-carbon retarded the cycloaddition.

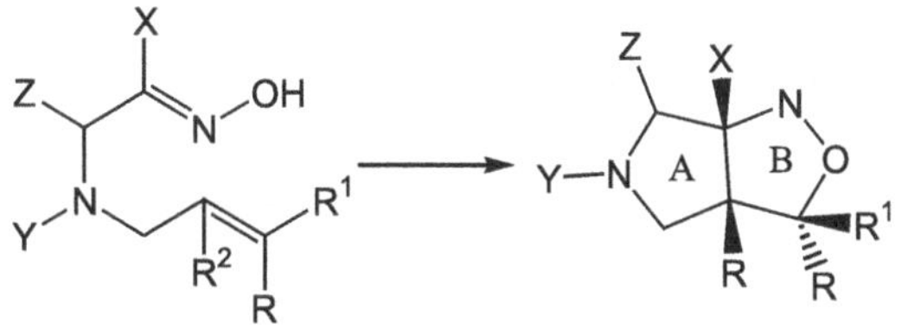

(12)

Calculations reveal that the presence of a methyl substituent R^1 adjacent to O in ring B and *syn* to the ring junction hydrogen prejudices a molecule in favor of conformer **71**, thus placing the methyl substituent pseudoequatorially. Similarly, a single substituent Z in the A ring, *syn* to the ring junction favors (**72**) (e.g., Z = Et) in which the A ring substituent Z can assume a pseudoequatorial position. The former effect dominates the latter when both rings are substituted affording the product with both side chains R^1 and Z in a β-orientation, but the preferred conformation is (**71**) and the Z group is forced into a pseudoaxial position.

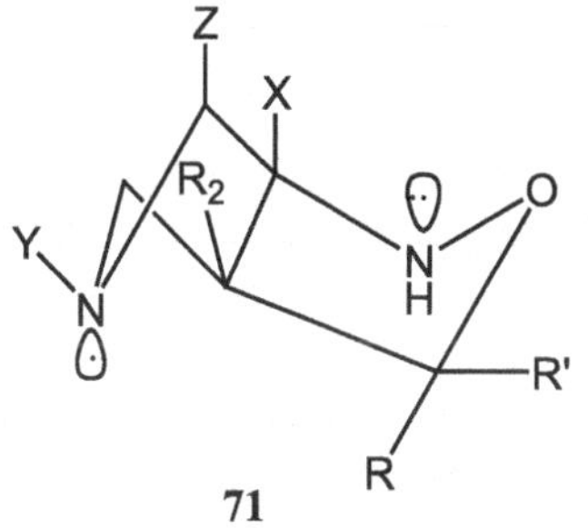

71

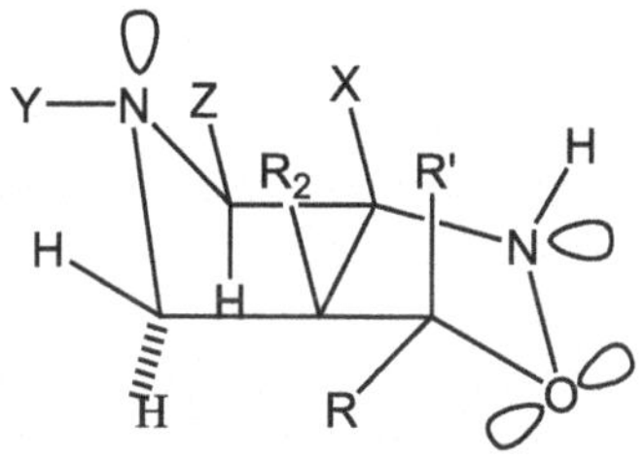

72

Hassner et al. also applied the IOOC reaction to the formation of carbocycles fused to an isoxazolidine. This required higher temperatures and was accomplished by heating the alkenyl oxime in a sealed tube at 190 °C [143]. The alkenyl oximes were prepared in good yield via reaction of carbanion (**73**) with *O*-silyl-α-bromoaldoxime in the presence of F^- ion. Reduction of the cycloadduct with LAH provides aminoalcohols in 68% yield (Scheme 67).

Formation of cyclohexane fused isoxazolidines from corresponding oximes by IOOC reaction was less selective, affording mixtures of isomers in 51%

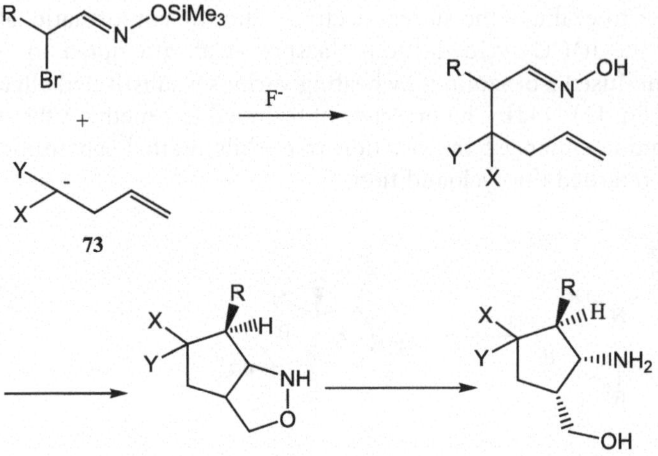

Scheme 67

yield. On the other hand, citronellal oxime underwent smooth cyclization when heated at 190 °C for 5 h to give the cycloadduct stereospecifically in 80% yield (Eq. 13).

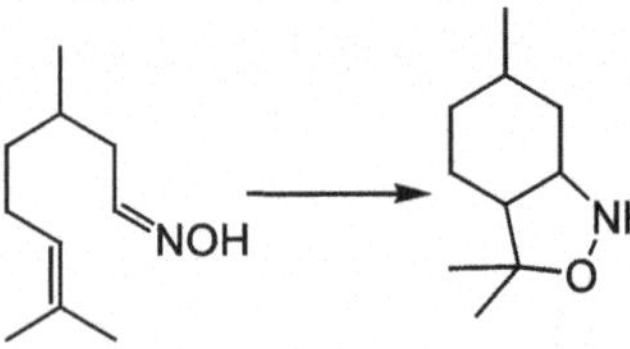

(13)

From the above, it is interesting to note that while the temperature required to effect IOOC of an oxime possessing an amine *N*-tether is only ca. 80 °C, an unsaturated oxime possessing an ether O required 110 °C, and the all carbon system proceeded at even higher temperature (190 °C). However it is not clear if the presence of the unshared electron pair on the amine N or the ether O exercise an assisting effect in the proton transfer from O to N in the starting oxime to produce the NH-nitrone [144].

Longer chain alkenylamine oximes likewise underwent thermally induced 1,3-dipolar cycloaddition under mild conditions leading to isoxazolo[4,3-c]-pyridine derivatives. The methyl groups at the 2-position of oximes (**74**) should facilitate the cycloaddition due to restriction of the conformational flexibility of the reaction sites and promotion of the isomerization to an NH-nitrone intermediate. The formation of *cis*-fused nitrone cycloadducts (**75**) was explained on the basis of transition state geometry. Consideration of Dreiding models suggested that the substituents at the 2-position of oximes (**74**) were essential and should prefer the *endo* transition state of the (E)-nitrone leading to *cis*-fused products (Scheme 68) [144].

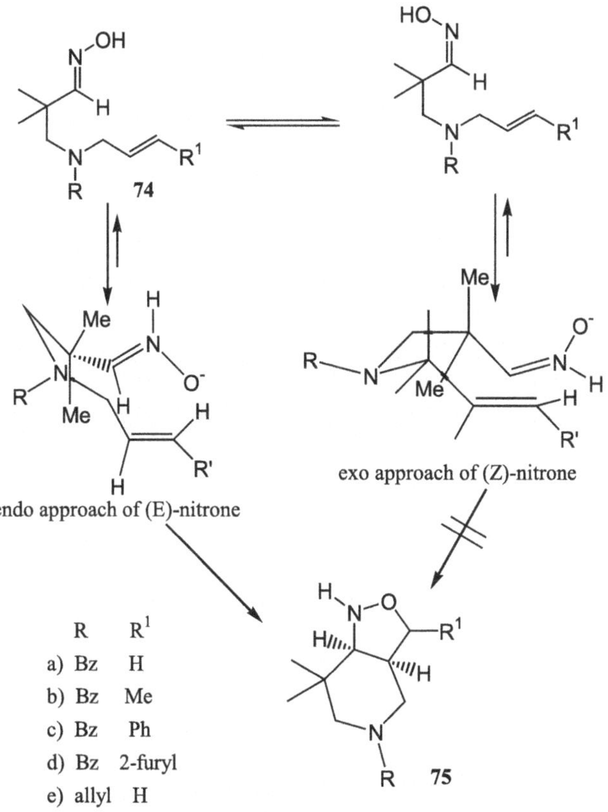

Scheme 68

Sammer and coworkers showed that sterically bulky groups adjacent to the nitrone dipole and the dipolarophile moieties could facilitate the rate of cycloaddition due to restriction of the conformational space of the dipole and dipolarophile [145].

In general, IOOC reactions have been shown to be convenient stereoselective routes not only to isoxazolidine derivatives but also for the stereoselective introduction of aminoalcohol functionality or of hydroxy ketones via N–O bond cleavage. Their scope extends to synthesis of carbocycles, as well as heterocycles bearing O, N or S atoms.

Shipman et al. utilized IOOC reactions as the key step for the stereocontrolled synthesis of polyhydroxylated aminocyclopentitols [146]. The starting 5-alkenyl aldehyde derived from various carbohydrates like D-glucose, D-mannose, D-galactose, D-glucal can be transformed into the corresponding oxime. Thermolysis of these oximes in toluene at 110 °C for 15 h afforded hexahydro-1*H*-cyclopent(c)isoxazole in good yield via intramolecular oxime-olefin cycloadditions (Scheme 69).

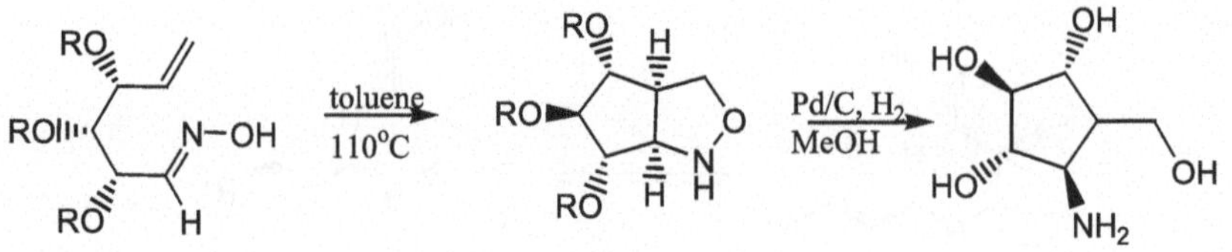

Scheme 69

Carbohydrate-derived alkenes have proved of use as key intermediates in the synthesis of a number of natural products. For instance, thermolysis in refluxing toluene for 15 h of an alkenyl oxime derived from carbohydrates yielded the fused isoxazolidine in quantitative yield (Scheme 70) [147]. It was suggested that the benzyl ethers were in a pseudoequatorial position in the transition state.

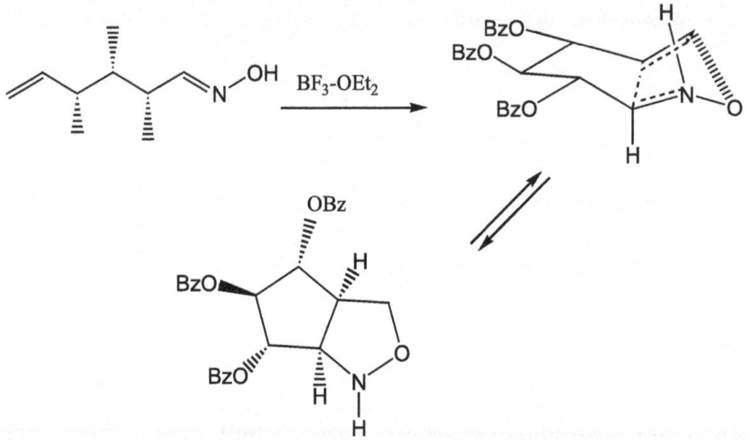

Scheme 70

When the oxime possesses a substituent (OBn or OBz) adjacent to the oxime carbon, there is a preference to produce the diastereomeric cycloadduct in which this substituent is located in an *exo* orientation. The alkenyl oximes derived from glucose, mannose, galactose afforded diastereomeric hexahydro-1*H*-cyclopent(C)isoxazolines (**76**), (**77**) and (**78**) respectively.

These isoxazolines can be transformed into stereochemically defined aminocyclopentitols. For instance, the selective mannosidase inhibitor mannostatin (**79**) and the very strong and specific trehalase inhibitor trihazoline (**80**) containing a polyhydroxylated aminocyclopentane moiety were synthesized via the intramolecular oxime cycloaddition strategy [146].

Hassner et al. have shown that substituted 5-membered aza-sugar analogues possessing an amino function can be synthesized stereospecifically from naturally occurring amino acids (e.g., L-serine) by an IOOC reaction (Scheme 71).

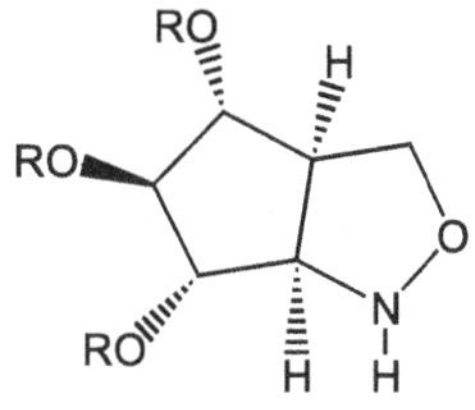

76

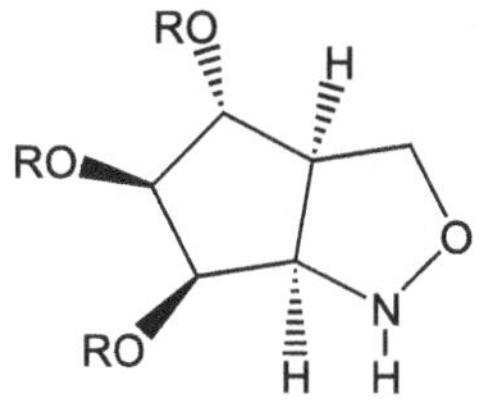

77

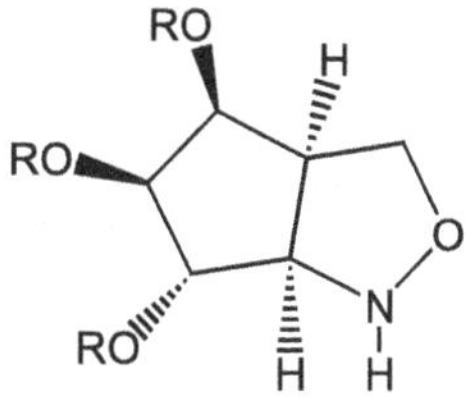

78

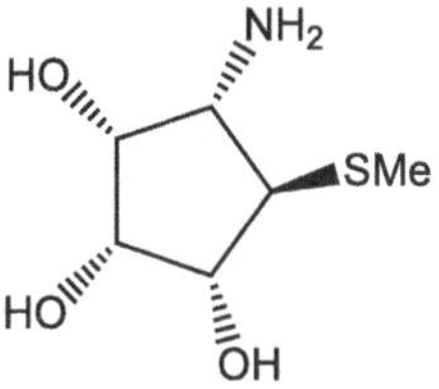

79

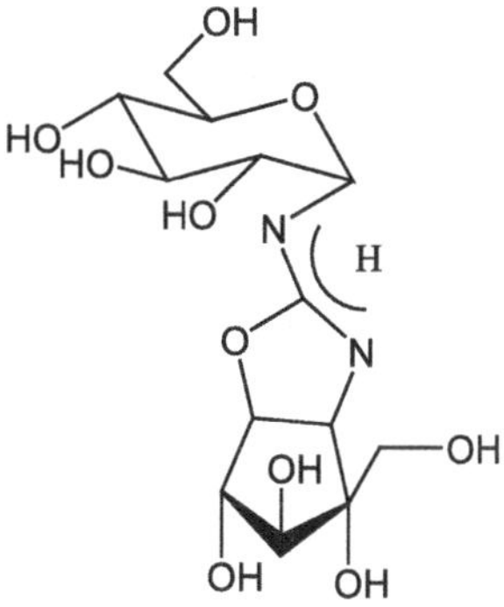

80

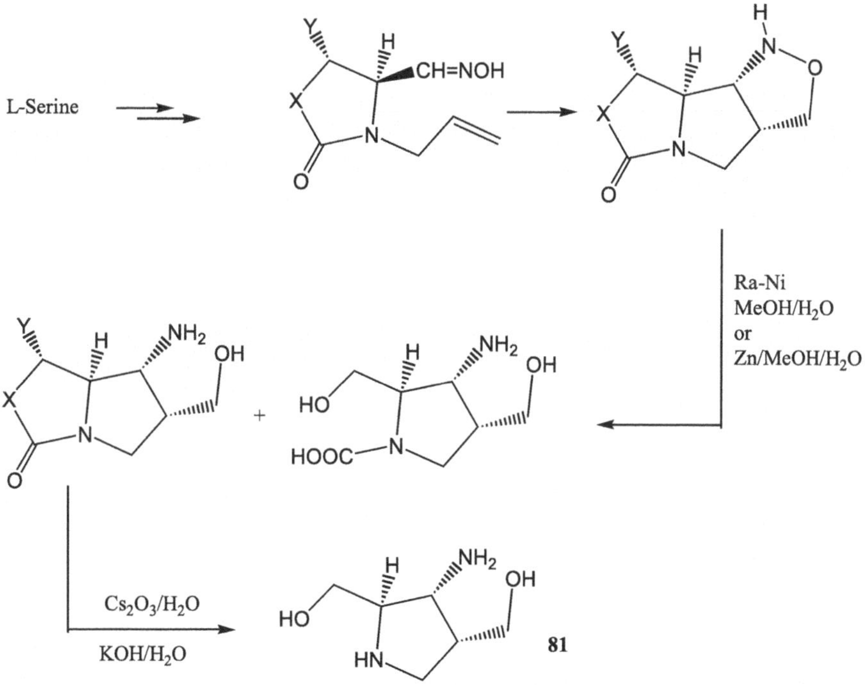

Scheme 71

This provided evidence for the first time that aza-sugars in which an OH was replaced by a NH_2 group (e.g., **81** in Scheme 71) possess glycosidase inhibitory properties [148]. The analog obtained from D-serine using microwave conditions was later found to be more active than the one obtained from L-serine [149].

Lewis acids may also catalyze the intramolecular cycloaddition of alkenyl oximes, which may proceed via an *N*-metallonitrone, to provide fused isoxazolidine derivatives (Scheme 72) [150].

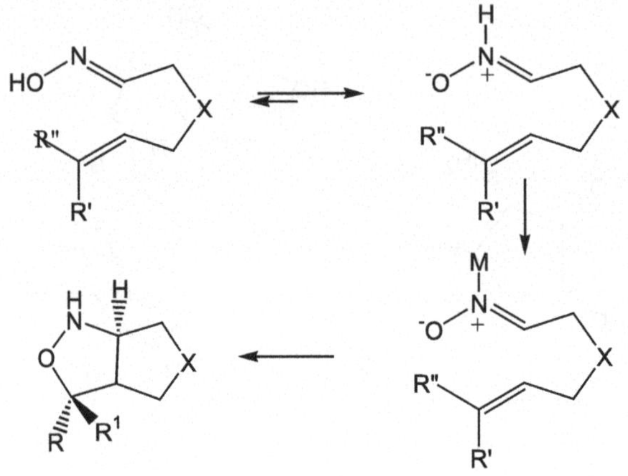

Scheme 72

O-tert-butyldimethylsilyloximes possessing an olefin moiety on treatment with 2 equivalents of BF_3–OEt_2 in CH_2Cl_2 undergo intramolecular cycloaddition at room temperature via *N*-boronitrones, to afford a cycloadduct in 99% yield (Scheme 73) [151].

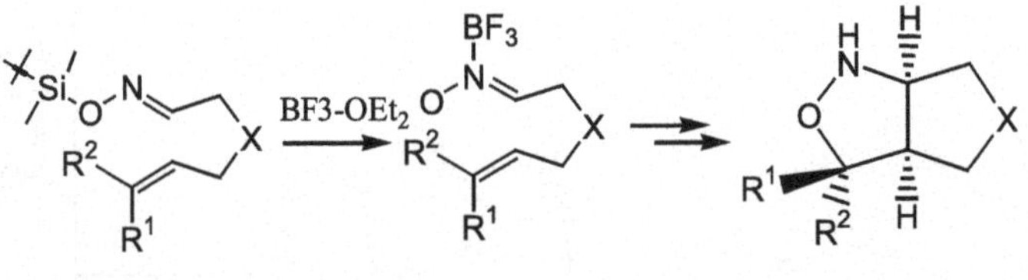

Scheme 73

1.7 [4+2] Cycloadditions of Nitrosoalkenes

α-Nitrosoalkenes are very useful synthetic intermediates possessing a double bond in conjugation with the nitroso group [152]. These reactive species along with nitrosocarbonyls constitute the two major sources for the construction of the 1,2-oxazine structure. Nitrosoalkenes are unstable and highly reactive. Normally they are observed only in solution, their presence sometimes being detectable by blue coloration (they have a λ_{max} close to 700 nm). Nitrosoalkenes have been isolated in only a few cases [153].

The formation of nitrosoalkenes, as fleeting intermediates, by reaction of *O*-silyl-α-bromooximes (**26**) with F^- and trapped in situ by unsaturated nucleophiles [97] has been described (Scheme 44). The usual method of generating nitrosoalkenes (**83**) is by the elimination of hydrogen halide from α-monohaloketoximes (**82**) in the presence of a base. The generated nitrosoalkenes are trapped by alkenes to produce 1,2-oxazine derivatives (**84**) (Scheme 74) [154].

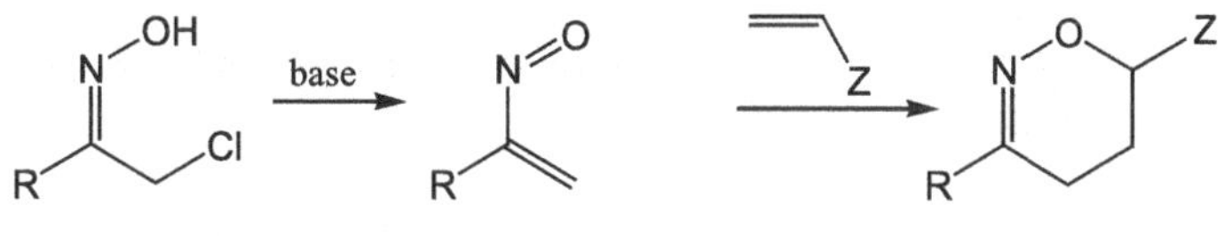

82 **83** **84**

Scheme 74

1,2-Oxazines (**87**) with a vinylic double bond at position 3 can easily be prepared by a hetero-Diels–Alder reaction of an electron-rich olefin (**86**) with the nitrosoalkene (**85**) generated in situ from methyl (E)-5-bromo-4-hydroximino-2-pentenoate and base by the procedure in Scheme 64 (Eq. 14) [155].

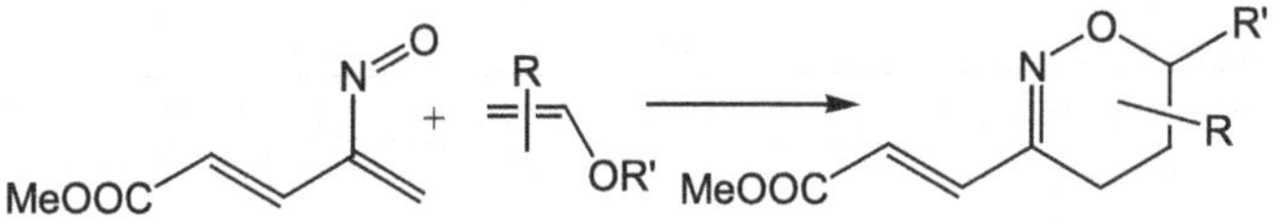

(14)

85 **86** **87**

Cyclization or cycloaddition of a nitrosoalkene (acting either as a 2π or a 4π component) with dienes, has been extensively covered by Gilchrist [156] and recently by Lyapkalo and Ioffe [157].

The cycloaddition of (**88**) with alkenes, in which (**88**) acts as a 4π conjugated heterodiene, is, perhaps, the most important reaction of nitrosoalkenes that has dominated their chemistry for quite a few years (Eq. 15) [158, 159].

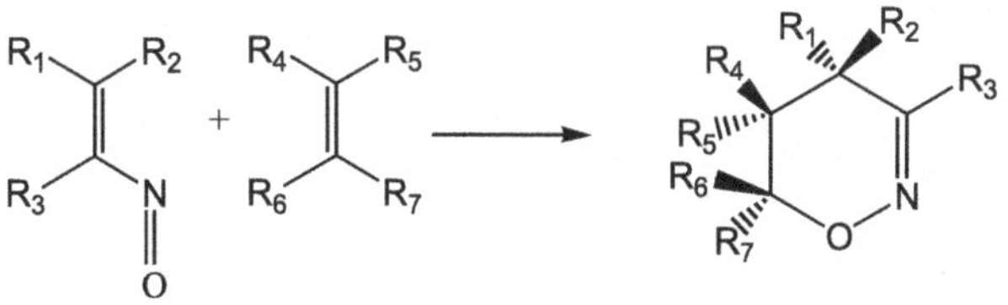

(15)

88 **89**

Allenes (**90**) upon reaction with nitrosoalkenes (**91**), gave the cycloadduct 1,2-oxazine (**92**) (Eq. 16).

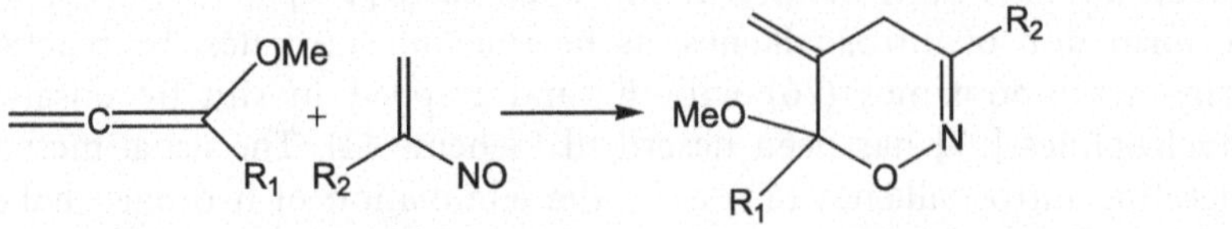

(16)

90 **91** **92**

Extending the [4+2] cycloaddition chemistry of nitrosoalkenes the reaction of 1,1,1-trifluoro-2-nitrosopropene (**93**) with silyl enol ethers (**94**) offers a flexible and efficient access to a large variety of 1,2 oxazines (**95**), which can be useful intermediates toward other functionalized compounds (Scheme 75) [160, 161].

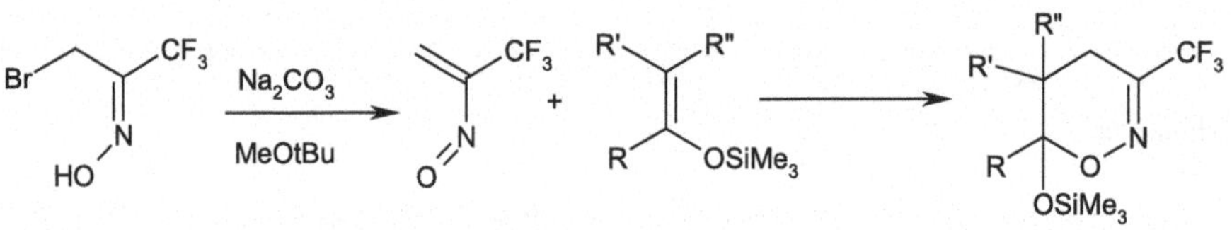

93 **94** **95**

Scheme 75

Ressig and coworkers [162] demonstrated that (**93**) is a powerful heterodiene, remarkably more reactive than nitrosostyrene or other alkyl bearing analogues, due to the electronic effects of the CF_3 group. Comparison of

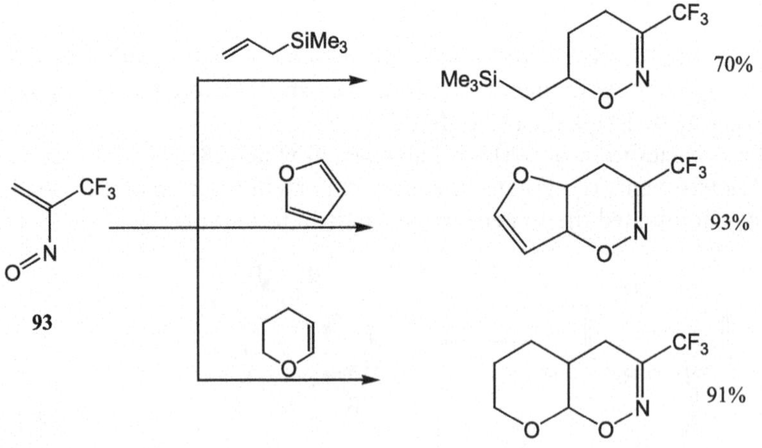

93

Scheme 76

HOMO and LUMO energies of (**81**) with those of nitrosoethylene or nitrosostyrene by MNDO calculations support the electronic accelerative effect of the group [162].

Dienophiles (**94**) bearing substituents larger than ethyl have been found to give no cycloadducts with nitrosostyrene. In contrast (**93**) has readily been added to unreactive silyl enol ethers in low to moderate yields. Other dienophiles have also been used and the corresponding 1,2-oxazines have been obtained in good to excellent yields (Scheme 76).

Recently Rai et al. utilized chloramine-T for the generation of nitrosoolefin starting from simple ketooximes. The resultant nitrosoolefins cycloadd with a dienophile leading to oxazine derivatives in almost quantitative yield (Scheme 77) [163].

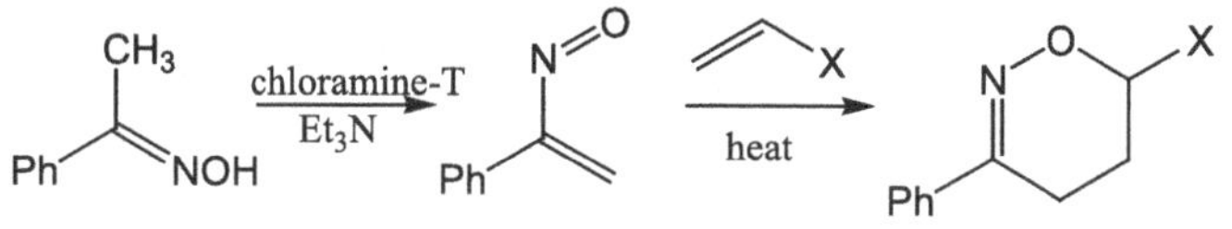

Scheme 77

1.7.1 Mechanism for the Generation of α-Nitrosoolefins from Ketoximes

When a ketoxime containing an α-methylene group was treated with chloramine-T a blue coloration was produced, suggestive of the formation of an α-chloronitroso compound [28], which on exposure to triethylamine generates α-nitrosoolefins (**99**). The latter, on treatment with an alkene, produced 1,2-oxazine derivatives. The formation of (**99**) from oxime (**96**) may be rationalized as shown in Scheme 78.

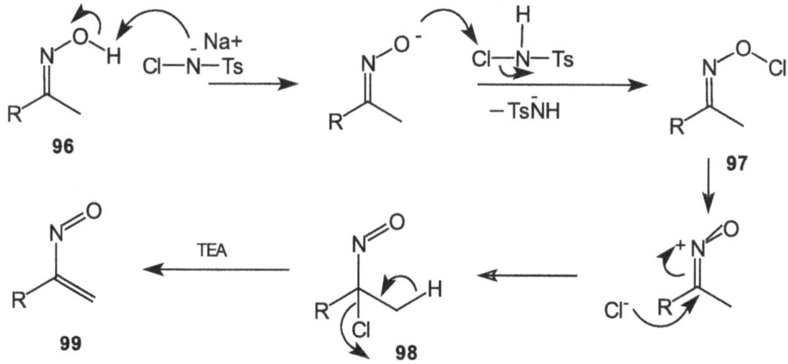

Scheme 78

In this scheme chloramine-T (or TEA) acts as a base, a well as a chlorinating agent transforming the oxime (**96**) into (**97**). The latter is slowly converted to α-chloronitroso compound (**98**), which leads to α-nitrosoolefin (**99**) on HCl elimination with TEA. This mechanism is consistent with a similar one proposed by Hassner and Rai for the generation of nitrile oxides from aldoximes using chloramine-T [28]. Alternatively, elimination of HCl from (**97**) may lead to (**99**).

1.8 Miscellaneous Reactions of Aldoximes

Padmavathi et al. observed an unusual reaction of oximes with excess chloramine-T [164]. For instance, treatment of *p*-chlorobenzaldoxime with 2 equivalent of chloramine-T in the presence of a dienone in refluxing methanol for 48 h afforded *N*-(4-tolyl)*N*-tosylbenzamide (**100**) instead of the

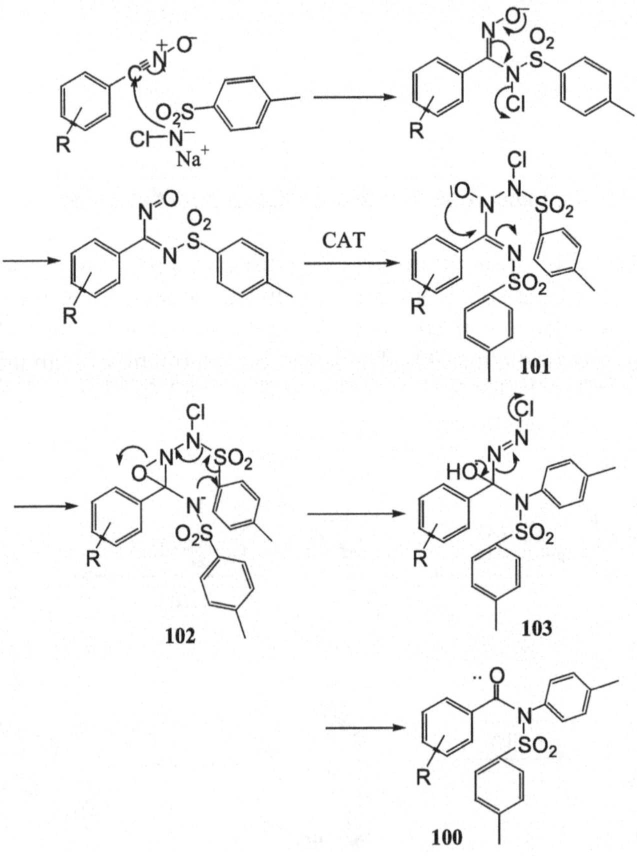

Scheme 79

usual addition product. When the reaction was carried out in the absence of olefin, they observed the formation of the same product. A tentative mechanism for the formation of (**100**) is depicted in Scheme 79. First, chloramine-T oxidizes the oxime to a nitrile oxide. Addition of another molecule of chloramine-T to the nitrile oxide results in the formation of a nitroso compound which is stabilized by the presence of the chloro substituent at the *ortho* position. In the absence of a stabilizing group at the *ortho* position, one more molecule of chloramine-T adds to the nitroso group to generate the intermediate (**101**), which undergoes an intramolecular rearrangement producing intermediate (**102**) bearing the oxaziridine ring system. Now the amide anion attacks the benzene ring to form chlorodiazo compound (**103**) with loss of SO_2. Finally loss of nitrogen and chloride ion transforms the intermediate (**103**) into the final *N*-(4-tolyl)benzamide (**100**).

References

1. Huisgen R (1963) Angew Chem Int Ed 2:565
2. Padwa A (1976) Angew Chem Int Ed 15:123
3. Padwa A (ed) (1984) 1,3-Dipolar Cycloaddition Chemistry. Wiley, New York
4. Padwa A (1991) In: Trost BM, Fleming I (eds) Comprehensive Organic Synthesis, vol 4. Pergamon, Oxford, p 1069
5. Wade PA (1991) In: Trost BM, Fleming I (eds) Comprehensive Organic Synthesis, vol 4. Pergamon, Oxford, p 1111
6. Gothelf KV, Jorgensen KA (1998) Chem Rev 98:863–909
7. Karlsson S, Högberg H-E (2001) Org Prep Proced Int 33:103
8. Kanemasa S (2002) Synlett, p 1371
9. Kanemasa S, Tsuge O (1990) Heterocycles 30:719
10. Gothelf KV (2002) Synthesis, p 211
11. Broggini G, Molteni G, Terraneo A, Zecchi G (2003) Heterocycles 59:823
12. Namboothiri INN, Hassner A (2001) Topics Curr Chem 216:1–49
13. Torssell KBG (1988) Nitrile oxides, nitrones and nitronsates in organic synthesis. VCH, Weinheim
14. Kim JN, Ryu EK (1990) Heterocycles 31:1693
15. Pellissier H (2007) Tetrahedron 63:3235
16. Just G, Dahl K (1968) Tetrahedron 24:5251
17. Mukaiyama T, Hosino T (1982) J Am Chem Soc 82:5339
18. Basel Y, Hassner A (1997) Synthesis 1997:309
19. Shimizu T, Hayashi Y, Shibafuchi H, Teramura K (1986) Bull Chem Soc Jpn 59:2827
20. Werner A, Buss H (1894) Ber Dsch Chem Ges 27:2193
21. Wieland H (1907) Ber Dsch Chem Ges 40:1667
22. Kotyatkina AI, Zhabinsky VN, Khripach VA (2001) Russ Chem Rev 70:641
23. Tokunaga Y, Ihara M, Fukumoto K (1996) Heterocycles 43:1771
24. Daiz-Ortiz A, Diez-Darra E, De la Hoz A, Moreno A, Gomez-Escalonilla E, Loupy A (1966) Heterocycles 43:1021
25. Grundmann C, Dean RJM (1965) J Org Chem 30:2809
26. Grundmann C, Richter R (1968) J Org Chem 33:476

27. Kim JN, Ryu EK (1990) Synth Commun 20:1373
28. Hassner A, Rai KML (1989) Synthesis 1989:57
29. Linganna N, Hassner A, Rai KML, Murthy CA (1992) Org Prep Proced Int 24:91
30. Gagneux AR, Meier R (1970) Helv Chim Acta 53:1883
31. Giurg M, Mlochowski M (1997) Polish J Chem 71:1093
32. Avai N, Iwakoski M, Tanaka K, Naranaka K (1999) Bull Chem Soc Jpn 72:2777
33. Radhakrishna AS, Sivaprakash K, Singh BB (1991) Synth Commun 21:1625
34. Moriya O, Takenaka H, Iyoda M, Urata Y, Endo T (1994) J Chem Soc Perkin Trans 1:413
35. Rai KML, Hassner A (1997) Synth Commun 27:467
36. Rai KML, Hassner A (1997) Ind J Chem Sec B 36B:242
37. Keigiel J, Poplawska M, Jozwik J, Kosior M, Jurczak J (1999) Tetrahedron 40:5605
38. Caramella P, Grunangar P (1984) In: Padwa A (ed) 1,3-Dipolar Cycloaddition Chemistry, vol 3. Wiley, New York, p 337
39. Dau S, Ducrosx B, Bernard S, Nicolas KM (1996) Tetrahedron Lett 37:4341
40. Umesha KB, Ajaykumar K, Rai KML (2002) Synth Commun 32:1584
41. Shang, YJ, Shou WG, Wang YG, (2003) Synlett, p 1064
42. Al-Dulayymi JR, Baird MS, Pavlov VA, Kurdjukov AI (1996) Tetrahedron Lett 52:8877
43. Houk KN, Sims J, Watts CR, Luskus LJ (1973) J Am Chem Soc 95:7301
44. Stork G, Danishefsky S, Ohashi M (1967) J Am Chem Soc 89:5459
45. Baraldi PG, Barco A, Benetti S, Manfredini S, Simoni D (1987) Synthesis 1987:276
46. Amici MD, Magri P, Micheli CP, Cateni F, Carrea R, Riva S, Caslone G (1992) J Org Chem 57:2825
47. Martin SF, Dappen MS, Dupré B, Murthy CJ, Colapret JA (1989) J Org Chem 54:2209
48. Feddouli A, Ait Itto MY, Hasanaoui A, Villemin D, Jaffris PA, Sopkova De OSI, Riachi A, Huel F, Daran JC (2004) J Heterocycl Chem 41:731
49. Neck TV, Mirtoo SV, Dehaen W (2007) Molecules 12:395
50. Zerressen A, Jaschke A (2005) J Mol Evol 61:236
51. Yadav PP, Ahmed G, Maurya R (2005) Tetrahedron Lett 46:5621
52. Ahmed G, Mishra PK, Gupta P, Yadav PP, Tiwari P, Tamrakar AK, Srivastava AK, Mourya R (2006) Bio Org Med Chem Lett 16:2139
53. Chai CLL, Edward AJ, Willer BA, Woodgate RCJ (2003) Tetrahedron 59:8731
54. Kikuchi K, Hiti S, Yoshimera H, Tokutana N, Tai K, Hide T, Yamauchi T, Nasai M (2000) J Med Chem 43:409
55. Simoni D, Roberti M, Invidiata FP, Randanin R, Baruchello R, Malagutti C, Mazzalli A, Roni M, Grimoudo S, Capone F, Dushonchet L, Meli M, Raimondi MV, Landino M, D'Alexandro N, Tolomeo M, Arindom D, Lu S, Ben Brook DM (2001) J Med Chem 44:2308
56. Zhang LH, Anzalone L, Ma P, Kauffman GS, Strorace L, Ward R (1996) Tetrahedron Lett 37:4455
57. Zierke T, Kuekenhoehner T, Frank J, Ammermann E, Lorenz G (1996) Ger Offen DE 41 013 723
58. Zierke T, Kuekenhoehner T, Frank J, Ammermann E, Lorenz G (1996) Chem Abstr 125:247801p
59. Nieolaides DN, Fylaktakidou KC, Litinas KE, Hadjipavtou LD (1996) J Heterocycl Chem 33:967
60. Cohen VL, Kleinman E (1995) PCT Int Appl WO 24 398
61. Cohen VL, Kleinman E (1996) Chem Abstr 124:117297
62. Kim HC, Seo MJ, Kim JK, Lee JD, No Z, Kim HR (2004) Bull Korean Chem Soc 25:133

63. Zhang H, Chan WH, Lee AWH, Xia PF, Wong WY (2004) Lett Org Chem 1:63
64. Cheng WC, Liu Y, Wong M, Olmstead MM, Lam KS, Kurth MJ (2002) J Org Chem 67:5673
65. Bardhan S, Schmitt DC, Porco JA (2006) Org Lett 8:927
66. Pirrung MC, Tumey LN, Raetz CRH, Jackman JE, Snehalatha K, McClerren AL, Fierke CA, Gantt SL, Rusche KM (2002) J Med Chem 45:4359
67. Bode JW, Fraefel N, Muri D, Carreira EM (2001) Angew Chem Int Ed Engl 40:2082
68. Kozikowski AP, Ghosh AK (1993) Tetrahedron Lett 24:2623
69. Padmavathi V, Venugopal Reddy K, Padmaja A, Bhaskar Reddy D (2003) Phosph Sulfur Silicon 178:171
70. Padmavathi V, Venugopal Reddy K, Padmaja A, Bhaskar Reddy D (2002) Synth Commun 32:1227
71. Padmavathi V, Jaganmohana Reddy B, Bhaskar Reddy D, Venkatasubbaiah K (2004) New J Chem 28:1479
72. Padmavathi V, Jaganmohana Reddy B, Chandramoul Redy B, Padmaja A (2005) Tetrahedron 61:2407
73. Ukaji Y, Sada K, Inomata K (1993) Chem Lett, p 1847
74. Ukaji Y, Inomata K (2000) Proceedings of the Fourth International Electronic Conference on Synthetic Organic Chemistry (EC SOC-4), 1–30 Sept 2000. MDPI, Basel, Switzerland, A0016
75. Tsuji M, Ukaji Y, Inomata K (2002) Chem Lett, p 1112
76. Ukaji Y, Inomata K (2003) Synlett, p 1075
77. Meyer AG, Easton CJ, Lincoln SF, Simpson GN (1977) Chem Commun, p 1517
78. Brinkmann Y, Madhushaw RJ, Jazzar R, Bernardinelli G, Kündig EP (2007) Tetrahedron 63:8413
79. Sibi MP, Itoh K, Jasperse CP (2004) J Am Chem Soc 126:5366
80. Kamimura A, Kaneko Y, Ohta A, Matsuura K, Fujimoto Y, Kakehi A, Kanemasa S (2002) Tetrahedron 58:9613
81. Gallos JK, Koftis TV, Koumbis AE, Moustsos V (1998) Synthesis 1998:1289
82. Gallos JK, Koftis TV, Koumbis A (2001) J Chem Soc Perkin Trans 1:415
83. Ferrer RI, Middeton S (1953) Chem Rev 93:2779
84. Jáger V, Müller I (1985) Tetrahedron 41:3519
85. Kim BH, Jeong EJ, Jung WH (1995) J Am Chem Soc 117:6390
86. Lee CW, Hwang GT, Kim BH (2000) Tetrahedron Lett 41:4177
87. Kanemasa S, Okuda K, Yamamoto H, Koga S (1992) Tetrahedron Lett 38:4095
88. Ma Z, Li L, Rupp M, Zhang S, Zhang X (2002) Org Lett 4:897
89. Lee CKY, Holmes AB, Al-Duri B, Leeke GA, Santos RCD, Seville JPK (2004) Chem Commun, p 2622
90. Hassner A, Rai KML, Dehaen W (1994) Synth Commun 24:1669
91. Hassner A, Murthy KSK (1987) Tetrahedron Lett 28:683
92. Hassner A, Murthy KSK (1987) Tetrahedron Lett 28:4097
93. Hassner A, Murthy KSK, Padwa A, Bullock WH, Stall PD (1988) J Org Chem 53:5063
94. Padwa A, Chiacchi U, Dean DC, Schoffstall AM, Hassner A, Murthy KSK (1988) Tetrahedron Lett 29:4169
95. Hassner A, Murthy KSK, Padwa A, Chiacchi U, Dean DC, Schoffstall AM (1989) J Org Chem 54:5277
96. Hassner A, Dehaen W (1990) J Org Chem 55:5505
97. Noguchi M, Tsukimoto A, Kadowaki A, Hikata J, Kakehi A (2007) Tetrahedron Lett 48:3539

98. Scott JP, Mullens PR, Brewer SE, Brands KMJ, Chilenski JR, Davies AJ, Gibb AD, Lieberman DR, Oliver SF, Dolling UH (2006) Org Biomol Chem 4:1806
99. Sengupta J, Mukhopadhyay R, Bhattacharjya A, Bhadbhade MM, Bhosekar GV (2005) J Org Chem 70:8579
100. Duclos O, Dureault A, Depezay JC (1992) Tetrahedron Lett 33:1059
101. Scott JP, Oliver SF, Brands KMJ, Brewer SE, Davies AJ, Gibb AD, Hands D, Keen SP, Sheen FJ, Reamer RA, Wilson RI, Dolling UH (2006) J Org Chem 71:3086
102. Shing TKM, Wong WF, Cheng HM, Kwok WS, So KH (2007) Org Lett 9:753
103. Mukhopadhyay R, Datta S, Chattoadhyay P, Bhattacharjya A, Patra A (1996) Ind J Chem 35B:1190
104. Gallos JK, Koumbis AE (2003) Curr Org Chem 7:397
105. McCarvey GJ, Mathys JA, Wilson KJ (1996) J Org Chem 61:5704
106. Takahashi T, Nakazawa M, Sakamoto Y, Houk KN (1993) Tetrahedron Lett 34:4075
107. Majumdar S, Mukhopadyay R, Bhattacharjya A (2000) Tetrahedron 56:8945
108. Pal A, Bhattacharjya A, Bhattacharjee A (1999) Synthesis 1999:1599
109. Pal A, Bhattacharjya A, Bhattacharjee A, Patra A (1999) Tetrahedron 45:4123
110. Hieoson AT, Jeffery J, Szczur N (1995) Tetrahedron Lett 36:7731
111. Smith AL, Pitsinos EN, Hwang C-K, Mizuno Y, Saimoto H, Scarlato CR, Suzuki T, Nicolaou KC (1993) J Am Chem Soc 114:3134
112. Nicolaou KC, Hummel CW, Nakada M, Shibayama K, Pitsinos EN, Saimoto H, Mizuno Y, Baldenius K-U, Smith AL (1993) J Am Chem Soc 115:7625
113. Kozikowski AP, Li C-S (1987) J Org Chem 52:3541
114. Ihara M, Tokunaga Y, Taniguchi N, Fukumoto K (1991) Tetrahedron 47:6635
115. Hassner A, Murthy KSK (1986) Tetrahedron Lett 27:1407
116. Murthy KSK, Hassner A (1991) Isr J Chem 31:239
117. Hassner A, Murthy KSK, Maurya R, Dehaen W, Friedman O (1994) J Heterocycl Chem 31:687
118. Das B, Mahender G, Holla H, Banerjee S (2005) Arkivoc, p 27
119. L'abbe G, Wuytswinkel, Dahean W (1995) Bull Soc Chim Belg 104:629
120. Fernandez-Mateos A, Coca GP, Rubio GR (2005) Tetrahedron 61:8699
121. Chiacchio U, Corsaro A, Rescifina A, Bkaithan M, Grassi G, Piperno A, Privitera T, Romeo G (2001) Tetrahedron 57:3425
122. Nakai K, Doi T, Takahashi T (2005) Synlett, p 866
123. Bode JW, Carreira EM (2005) J Org Chem 66:6410
124. Rao KR, Bhanumathi N, Sattur PB (1990) Tetrahedron Lett 31:3201
125. Aunger RA, Chan C, Funk RL (2001) Org Lett 3:2611
126. Tatsuka K, Niwata Y, Umezava K, Toshima K, Nokala M (1990) Tetrahedron Lett 31:1131
127. Tatsuta K (1999) Pure Appl Chem 68:1341–1346
128. Shing TKM, Wong C–H, Yip T (1996) Tetrahedron Assymmetry 7:1323
129. Konojia RM, Chin E, Smith C, Chen R, Rowand D, Levine SD, Wachter MP, Adams RE, Hahn DW (1985) J Med Chem 28:796
130. Heinz I, Eberbach W (1988) Tetrahedron Lett, p 2051
131. Hassner A, Dehaen W (1991) J Org Chem 56:896
132. Hassner A, Murthy KSK (1987) Tetrahedron Lett 28:4097
133. Hassner A, Murthy KSK, Padwa A, Bullock WH, Stull PD (1988) J Org Chem 53:5063
134. Ferwanah ARS, Awadallah AM (2005) Molecules 10:492
135. Grigg R, Gunaraline HQN, Kemp J (1984) J Chem Soc Perkin 1:41
136. Grigg R (1987) Chem Soc Rev 16:89

137. Hassner A, Maurya R (1989) Tetrahedron Lett 30:2289
138. Hassner A, Maurya R, Padwa A, Bullock VH (1991) J Org Chem 56:2775
139. Hassner A, Falb E, Nudlemann A, Albeck A, Gotlieb HE (1994) Tetrahedron Lett 35:2393
140. Hassner A, Shing S, Sharma R, Maurya R (1993) Tetrahedron 49:2317
141. Hassner A, Maurya R, Mesko E (1988) Tetrahedron Lett 29:5313
142. Hassner A, Maurya R, Friedman O, Gottlieb HE, Padwa A, Austin D (1993) J Org Chem 58:4539
143. Hassner A, Maurya R (1989) Tetrahedron Lett 30:5803
144. Noguchi M, Okada H, Nishimura S, Yamagata Y, Takamura S, Tanaka M, Kekehi A, Yamamoto H (1999) J Chem Soc Perkin 1:185
145. Sammer PG, Weller DJ (1995) Synthesis 1995:1805
146. Dransfield PJ, Moutel S, Shipman M, Sik V (1999) J Chem Soc Perkin 1:3349
147. Moutel S, Shipman M (1988) Synlett, p 1333
148. Falb E, Bechor Y, Nudleman A, Hassner A, Albeck A, Gotlieb HE (1999) J Org Chem 64:498
149. Cheng Q, Zhang W, Tagami V, Oritani T (2001) J Chem Soc Perkin 1:452
150. Frederickson M, Grigg R, Thomton-Pet M, Redpath J (1997) Tetrahedron Lett 38:7777
151. Tamura O, Mitsuya T, Ishibashi H (2002) Chem Commun, p 1128
152. Gallos K, Sarli TV, Argyropoulou EC (2000) Tetrahedron Lett 41:4819
153. Deady LW (1967) Tetrahedron 23:3505
154. Gilchrist L, Roberts TG (1983) J Chem Soc Perkin Trans 1:1283
155. Zimmer R, Collas M, Roth M, Liebigs H-U (1992) Ann Chem, p 709
156. Gilchrist TL (1983) Chem Soc Rev 12:53
157. Lyapkalo IM, Ioffe S (1998) Russ Chem Rev 67:467
158. Davis DE, Gilchrist TL, Roberts TG (1983) J Chem Soc Perkin Trans 1:275
159. Ioffe SL, Lyapkalo IM, Tishkov AA, Danilenko VM, Strelenko YA, Tartakovsky VA (1997) Tetrahedron 53:13085
160. Hippeli C, Ressig H-U (1987) Synthesis 1987:77
161. Zimmer R, Ressig H-U (1988) Angew Chem Int Ed Engl 27:1518
162. Zimmer R, Ressig H-U (1992) J Org Chem 57:339
163. Gaonkar SL, Rai KML (2005) J Heterocycl Chem 42:877
164. Padmavathi V, Venugopal Reddy K, Padmaja A, Venugopala P (2002) J Org Chem 68:1567

Top Heterocycl Chem (2008) 13: 71–98
DOI 10.1007/7081_2007_093

Published online: 5 December 2007

Heterocycles via Pyrylium and Pyridinium Betaines

K. V. Radhakrishnan

Organic Chemistry Section, National Institute for Interdisciplinary Science and Technology (CSIR), 695 019 Trivandrum, India
radhupreethi@rediffmail.com

Abstract The applications of pyrylium and pyridinium betaines in the synthesis of heterocyclic systems are reviewed. The cycloadditions of pyrylium and pyridinium betaines offers an easy access to novel heterocyclic molecules with a nitrogen or oxygen bridge. These cycloadducts can be synthetically manipulated easily towards a number of interesting molecular structures with potential applications. It is to be noted that only the progress in this area made in the last 10 years is emphasized.

Keywords Dipolar cycloadditions · Heterocycles · Tropolones

Abbreviations

Ac	Acetyl
Bn	Benzyl
DBN	Diazabicyclo[4.3.0]non-5-ene
DBU	1,8-Diazabicyclo[5.4.0]undec-7-ene
DMAD	Dimethyl acetylenedicarboxylate
DMAP	Dimethylaminopyridine
Ms	Methane sulfonyl
NBS	*N*-Bromo succinimide
*i*Pr	Isopropyl

rt	Room temperature
Tf	Triflyl (trifluoromethane sulfonyl)
TMS	Trimethyl silyl
THF	Tetrahydrofuran
TBAF	Tetra-*n*-butyl ammonium fluoride
TFA	Trifluoro acetic acid
TBS	*t*-Butyl dimethylsilyl
t	Tertiary
TMP	Tetramethyl piperidine

1
Introduction

Fused ring heterocycles and carbocycles are found in structurally complex natural products possessing diverse biological activities and are valuable in accessing novel entities for the pharmaceutical industry. The development of elegant and clear-cut routes toward these fabulous targets has always elicited widespread interest in organic synthesis. In this perspective, dipolar cycloaddition reactions, with their synthetic brevity and true efficiency, have come into focus [1]. The versatility of dipoles and dipolarophiles and the scope of further transformation of the cycloadducts have made their chemistry an interesting approach, not only for simple fused heterocycles but also for the synthesis of complex natural products [2, 3].

1,3-Dipolar cycloaddition reactions have received considerable attention as they offer a convenient route to multifunctional compounds in a highly regiospecific and stereospecific fashion [4]. The structurally diverse groups of heterocyclic compounds that have been synthesized by 1,3-dipolar cycloadditions clearly attest their potential value in organic synthesis. Besides these, there are certain higher-order dipolar cycloadditions that provide access to medium-sized heterocyclic ring systems. Among the various higher-order dipoles available, 3-oxidopyrylium betaines and 3-oxidopyridinium betaines have attracted considerable interest. 3-Oxidopyrylium betaines and 3-oxidopyridinium betaines offer great synthetic value due to their potential ability to participate as 1,3-, 1,5- and 1,7-dipoles in cycloaddition reactions [5].

2
Generation of Pyrylium Betaines

Pyranulose acetates, precursors for the corresponding 3-oxidopyrylium betaines, were synthesized from furfuryl alcohol by oxidation with *N*-bromosuccinimide (NBS) (Achmatowiscz reaction) followed by acetylation. The 3-oxidopyrylium ylides can be generated from pyranulose acetate by means

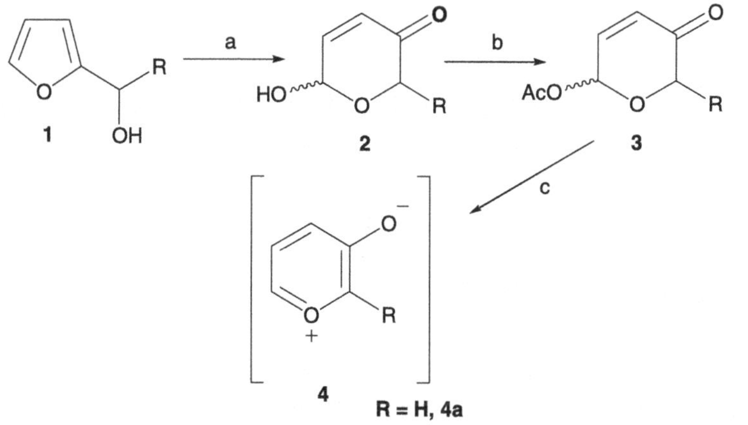

Scheme 1 Reagents and conditions: *a* NBS, THF : H_2O; *b* Ac_2O, pyridine, DMAP, CH_2Cl_2; *c* Et_3N or DBU

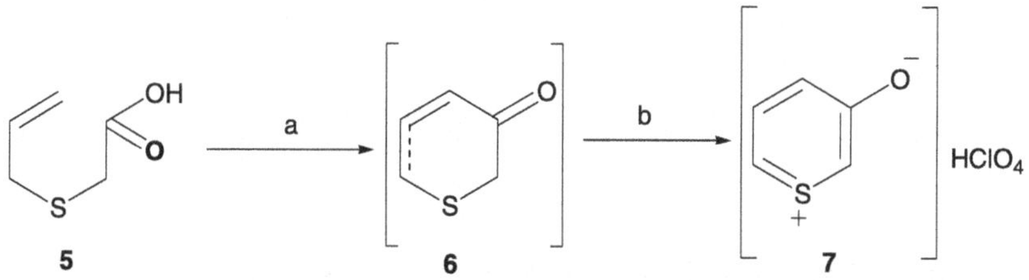

Scheme 2 Reagents and conditions: *a* $AlCl_3$; *b* triphenylmethyl perchlorate

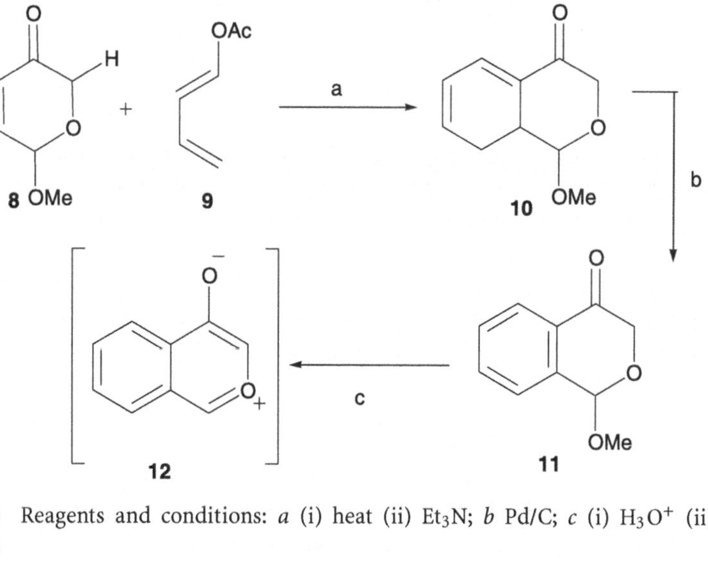

Scheme 3 Reagents and conditions: *a* (i) heat (ii) Et_3N; *b* Pd/C; *c* (i) H_3O^+ (ii) Ac_2O (iii) heat

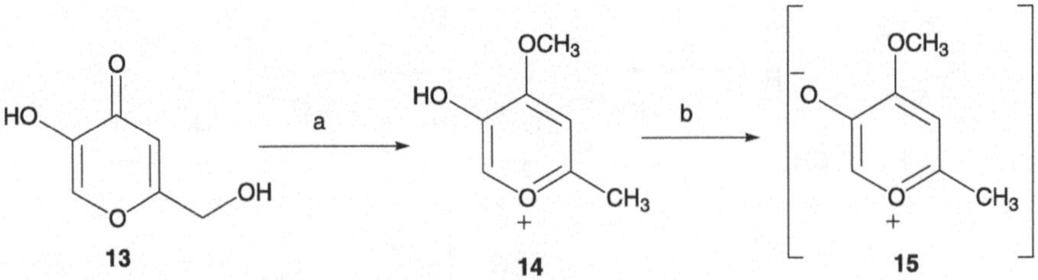

Scheme 4 Reagents and conditions: *a* (i) $SOCl_2$, $CHCl_3$, reflux (ii) H_2, Pd/C, NaOAc, MeOH; *b* MeOTf, CH_2Cl_2, reflux, 4 h

of a base such as triethylamine, diazabicyclo[4.3.0]non-5-ene (DBN) or DBU, or thermally in acetonitrile in a sealed tube at 150 °C (Scheme 1) [6, 7].

The sulfur analog of 3-oxidopyrylium ylide, namely 3-oxidothiopyrylium perchlorate **7**, is prepared from allylthioglycoloyl chloride as is shown in Scheme 2 [8].

4-Oxido-2-benzopyrylium betaine is conveniently prepared from the corresponding benzoannelated pyranoside **10** (Scheme 3) [8].

4-Methoxy-6-methyl-3-oxidopyrylium betaines (e.g., **15**) are readily generated from kojic acid **13** and an example is shown in Scheme 4 [9].

3 Cycloaddition Reactions of 3-Oxidopyrylium Betaines

3.1 With 2π Components – Intermolecular Mode

Cycloaddition reactions based on 3-oxidopyrylium betaines with alkenes have emerged as an efficient method of constructing seven-membered ring units. The pioneering works in the area were attributed to Hendrickson, for initiating the cycloaddition chemistry of pyrylium betaines [6, 7]. A number of electrophilic alkenes have been found to undergo [3 + 2] cycloaddition with pyrylium betaines producing oxabicyclo [3.2.1] systems, thus presenting a novel route towards seven-membered carbocycles (Scheme 5). The 3-oxidopyrylium betaine can be considered as a 4π-electron 1,3-dipole across the 2,6-positions, a 6π-electron 1,5-dipole across the 2,4-positions, and as an 8π-electron 1,7-dipole across O and C-2 or O and C-4. Additions of 2π addends are thermally allowed across the 2,6-positions and across the C-4, O and C-2, O termini. However, the preferred regioisomers with unsymmetrical alkenes have the 2,2′-6,1′ structure.

In an interesting extension of the above cycloaddition, Sammes and coworkers have shown that pyrylium betaines undergo cycloaddition with both electron-rich and electron-deficient alkenes (Scheme 6) [10]. The resultant

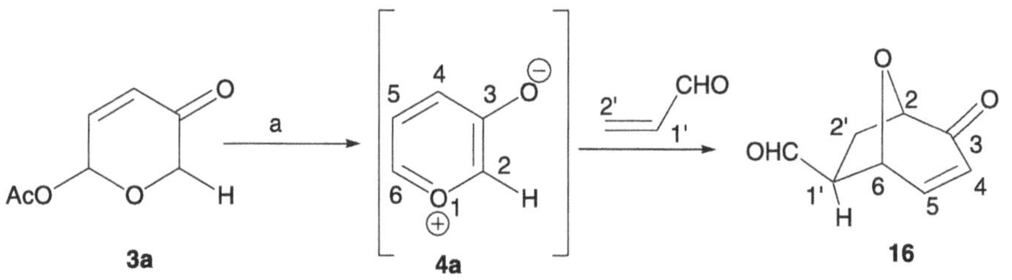

Scheme 5 Reagents and conditions: *a* heat, 69%

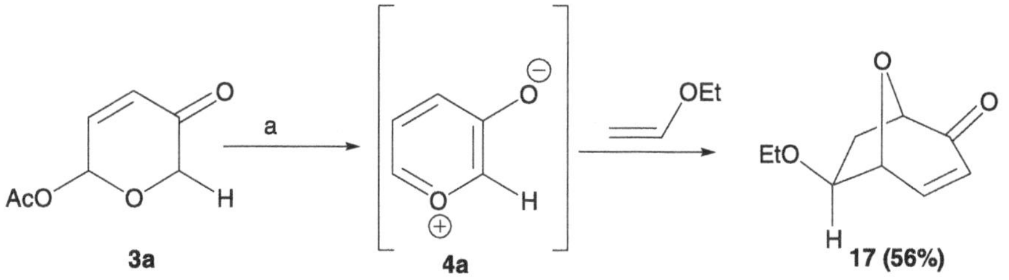

Scheme 6 Reagents and conditions: *a* Et_3N, rt

compounds have been shown to be excellent intermediates for the synthesis of potent herbicides and plant growth regulators.

Tropolone natural products are attractive targets in organic synthesis due to the unique structure and properties of the seven-membered tropolone ring. In this context, Baldwin and coworkers have developed an expedient synthesis of a substituted tropolone (i.e., **23**) which involves a 3-oxidopyrylium [5 + 2] cycloaddition with acrylonitrile as the key step (Scheme 7) [11].

A novel tropolone *ortho*-quinone methide precursor **27** for the deoxy analog of the natural product epolone B was synthesized through an oxidopyrylium betaine cycloaddition strategy and is shown in Scheme 8 [12].

One of the most synthetically useful aspects of concerted cycloaddition is the possibility of controlling the stereochemistry at a range of centers in the resulting cycloadducts. Utilizing this aspect, C.W.G. Fishwick et al. have reported a simple and efficient access to stereodefined tetrahydrofurans through reduction followed by ozonolytic cleavage of the [3 + 2] cycloadducts of 3-oxidopyrylium betaine to various olefins. An illustrative example is shown in Scheme 9 [13].

Heathcock and coworkers have developed a method for the synthesis of the 2,7-dioxatricyclo[4.2.1.0]nonane ring system characteristic of the marine diterpene dictyoxetane **35** [14] The method utilizes dipolar cycloaddition of a 3-oxidopyrylium betaine to an olefin to create the carbon skeleton. The developed route was easily adapted to incorporate additional functionality, making it useful in a total synthesis of dictyoxetane (Scheme 10) [14].

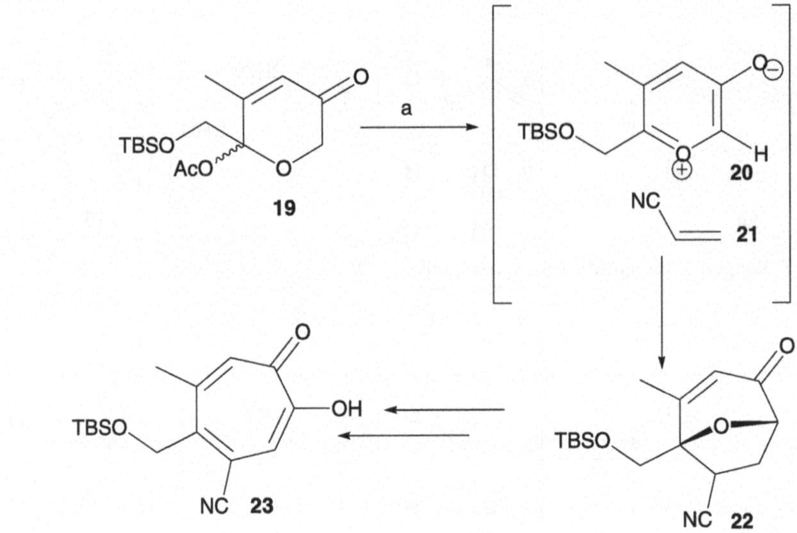

Scheme 7 Reagents and conditions: *a* toluene, 120 °C, 6 h, 42%

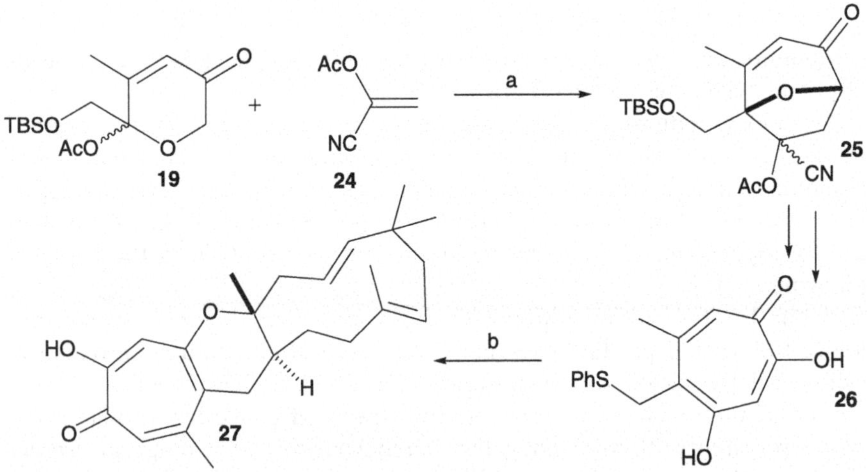

Scheme 8 Reagents and conditions: *a* toluene, 120 °C, 54%; *b* humulene, *p*-xylene, 150 °C

V.K. Aggarwal and coworkers have reported a diastereoselective 1,3-dipolar cycloaddition reaction of 3-oxidopyrylium betaine with a C2-symmetric vinyl sulfoxide, *trans*-2-methylene-1,3-dithiolane 1,3-dioxide (Scheme 11) [15].

G.K. Trivedi and coworkers reported synthetic studies towards the diterpene natural product, FCRR toxin **43**. The strategy utilizes an intermolecular [3 + 2] oxidopyrylium–indene cycloaddition and the reaction proceeded with very high regio- and stereoselectivity (Scheme 12) [16].

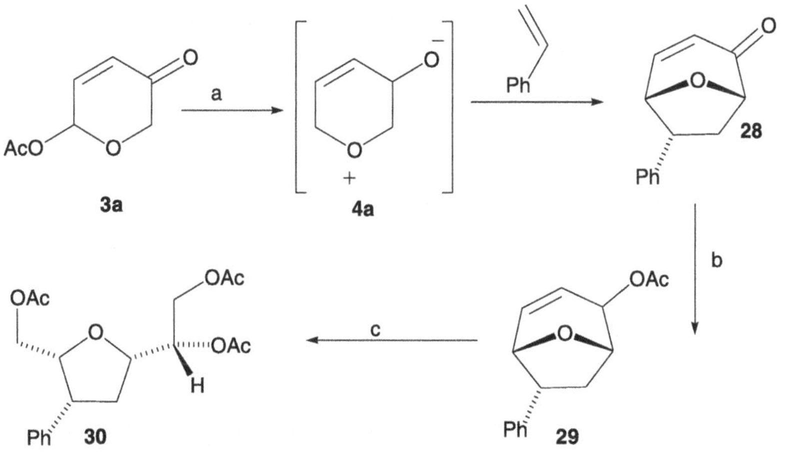

Scheme 9 Reagents and conditions: *a* Et_3N, rt, 79%; *b* (i) $NaBH_4$, $CeCl_3$ (ii) Ac_2O, pyridine; *c* (i) O_3, MeOH (ii) $LiAlH_4$ (iii) Ac_2O, pyridine (62%)

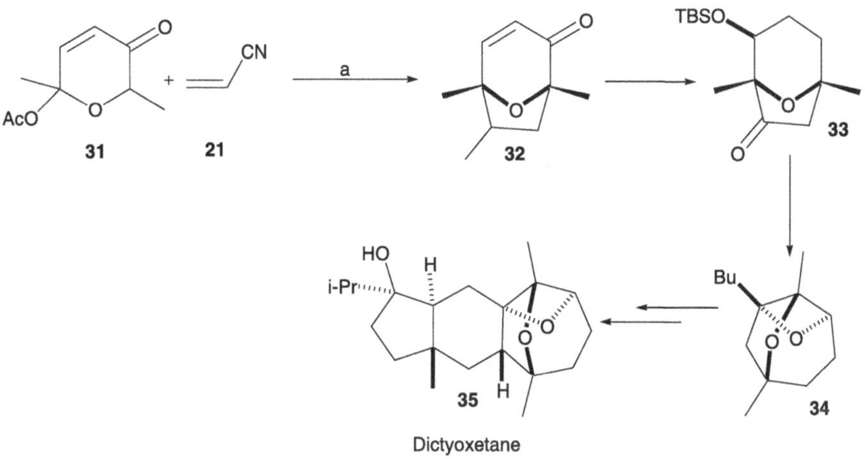

Scheme 10 Reagents and conditions: *a* MsCl, $(i\text{Pr})_2$EtN, CH_3CN, reflux

Snider and team mates have utilized the oxidopyrylium betaine [5 + 2] cycloaddition strategy towards the synthesis of cartorimine [17] and descurainin, utilizing the [3 + 2] cycloaddition reaction of oxidopyrylium betaine with a variety of styrenes. Heating the betaine precursor pyranulose **44** with styrenes **45** and **48** in presence of a base afforded the [3 + 2] adducts **46** and **47**, respectively, which were hydrolyzed to the natural products cartorimine and descurainin (Scheme 13) [17, 18].

Wender et al. reported the preparation of 4-methoxy-3-oxidopyrylium ylide **52** from 3-hydroxy-4-pyrones and its unprecedented use as a 5-carbon

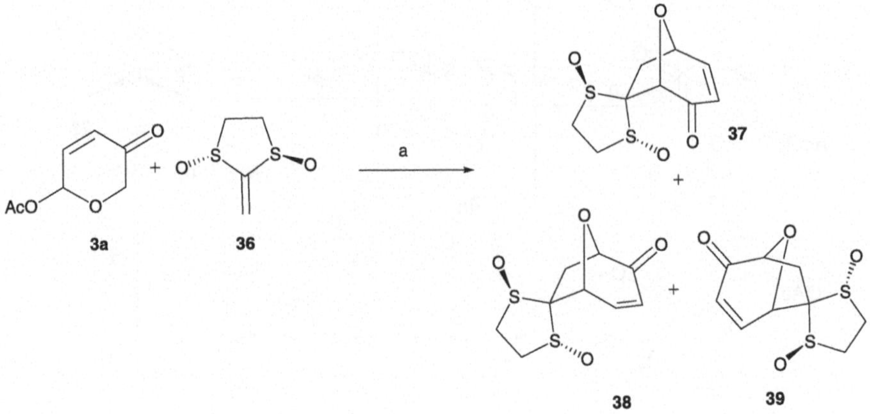

Scheme 11 Reagents and conditions: *a* iPr_2Net, CH_2Cl_2, 18 h, rt, 44%

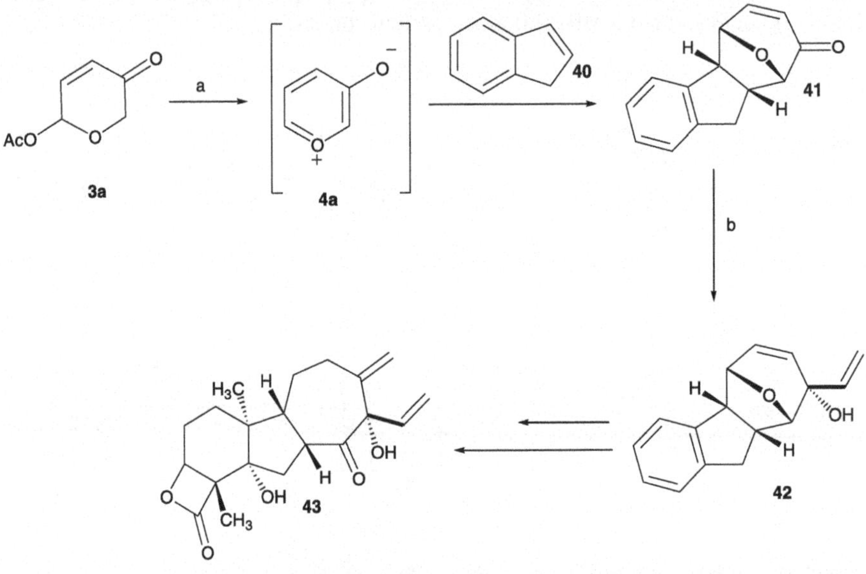

Scheme 12 Reagents and conditions: *a* Et_3N, CH_2Cl_2, 0 °C; *b* (i) H_2, Pd/C (10%) EtOH, 98% (ii) vinylmagnesium bromide (1 M in THF), −20 °C, 86%

dipole or 3-atom heterodipole reagent in [5 + 2] or [3 + 2] cycloadditions with various dipolarophiles; one such example is given in Scheme 14 [19].

The stereoselectivity of [3 + 2] cycloaddition of 2-substituted-3-oxidopyrylium betaines with dialkyl fumarates was found to depend on the bulkiness of the 2-substituent of 3-oxidopyrylium betaine and on the alkoxy group of the fumarate (Scheme 15) [20].

Lee and coworkers reported the cycloaddition reaction of 3-oxidopyrylium betaines with electron-rich allenes (Scheme 16). In this case, the cycloaddi-

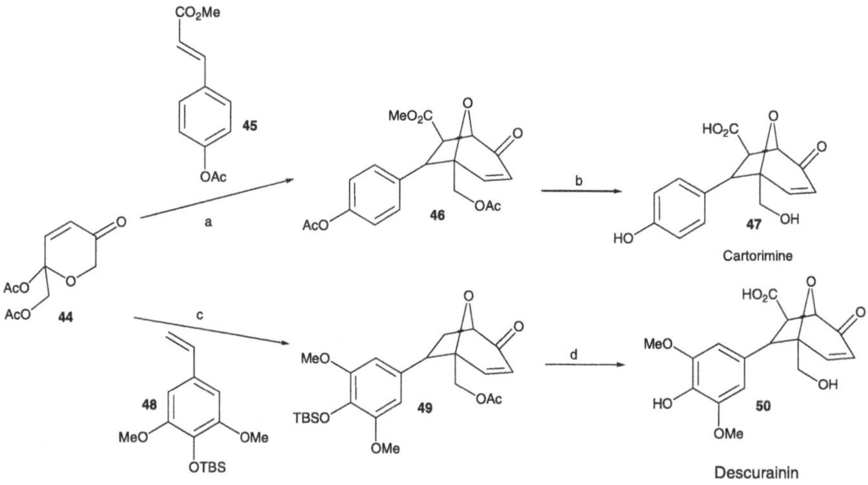

Scheme 13 Reagents and conditions: *a* 2,8-di-*t*-butylpyridine, CH_3CN, 125 °C; *b* KOH, EtOH; *c* Et_3N, CH_2Cl_2, 25 °C, 31%; *d* (i) pyridine.HF (ii) KOH, MeOH

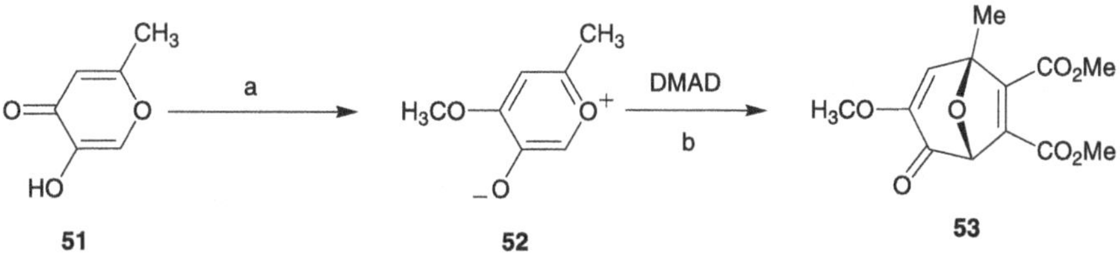

Scheme 14 Reagents and conditions: *a* MeOTf (1.5 equiv), CH_2Cl_2, reflux, 4 h; *b* 2,4,6-trimethyl pyridine, CH_2Cl_2, 73%

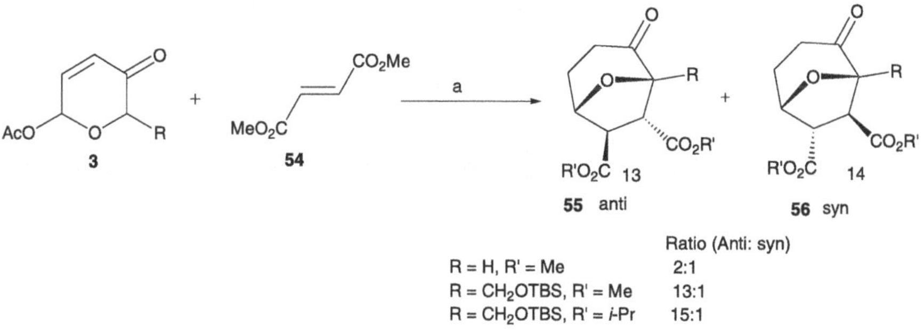

Scheme 15 Reagents and conditions: *a* Et_3N, rt

tion took place preferentially at the terminal position of the allene, regardless of the substitution pattern. However, electron-deficient allenes are found to be unreactive and electron-rich allenes were slightly more reactive compared to neutral allenes [21].

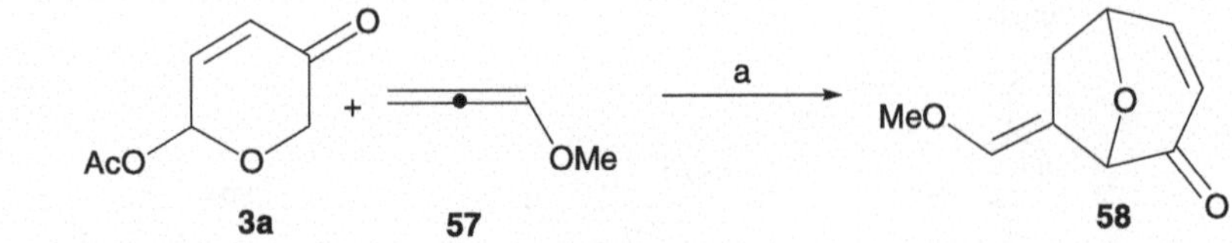

Scheme 16 Reagents and conditions: *a* Et_3N, rt, 46%

The reactions discussed above clearly indicate that 3-oxidopyrylium betaines are well known precursors for the synthesis of seven-membered rings. Their use in the synthesis of other medium-sized carbocycles has not been extensively studied. However, there are a few isolated reports on the formation of eight-membered rings via pyrylium betaine-mediated cycloadditions.

Mascarenas and his team have disclosed a [5C + 2C] oxidopyrylium–cyclopropenone acetal cycloaddition and a subsequent ring opening of the cyclopropane unit of the resulting adduct, leading to highly substituted 1,5-oxabridged cyclooctenes (see **61**), compounds of potential synthetic value that are not easily accessible through available procedures (Scheme 17) [22].

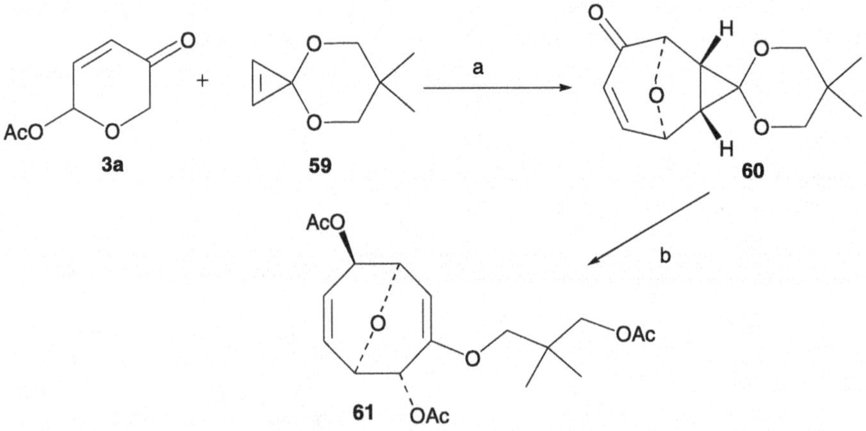

Scheme 17 Reagents and conditions: *a* Et_3N, CH_2Cl_2; *b* (i) $NaBH_4$ (ii) Ac_2O, TMSOTf

3.2
2π Intramolecular Mode

An asymmetric version of oxidopyrylium–alkene cycloaddition, leading to the enantiopure synthesis of oxabicyclo[5.4.0]undecanes was reported by G.K. Trivedi and coworkers. Optically active unsaturated aldehydes, derivable from the appropriate sugar derivatives, on reaction with 2-lithio furan provides useful 2-furyl carbinol intermediates. Conversion of the carbinol to pyranulose acetate, followed by treatment with a base afforded oxabicyclo [5.4.0]undecanes **66** in good yield. An example is shown in Scheme 18 [23].

J.L. Mascarenas et al. reported that the cycloaddition between 3-siloxy-4-pyrones and alkenes proceeded with complete regio- and stereoselectivity

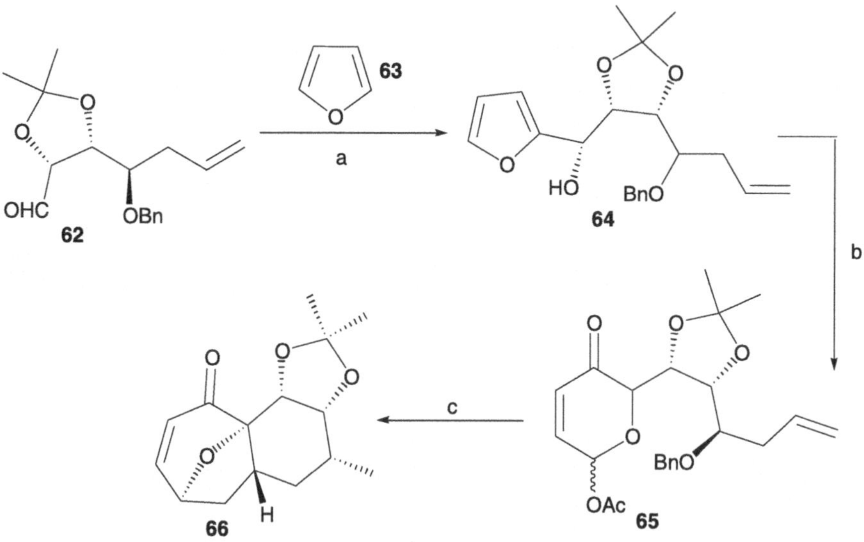

Scheme 18 Reagents and conditions: *a* *n*-BuLi, THF, –78 °C; *b* (i) $VO(acac)_2$, *t*-BuOOH (ii) Ac_2O, pyridine, 88%; *c* CH_3CN, reflux, Et_3N, 65%

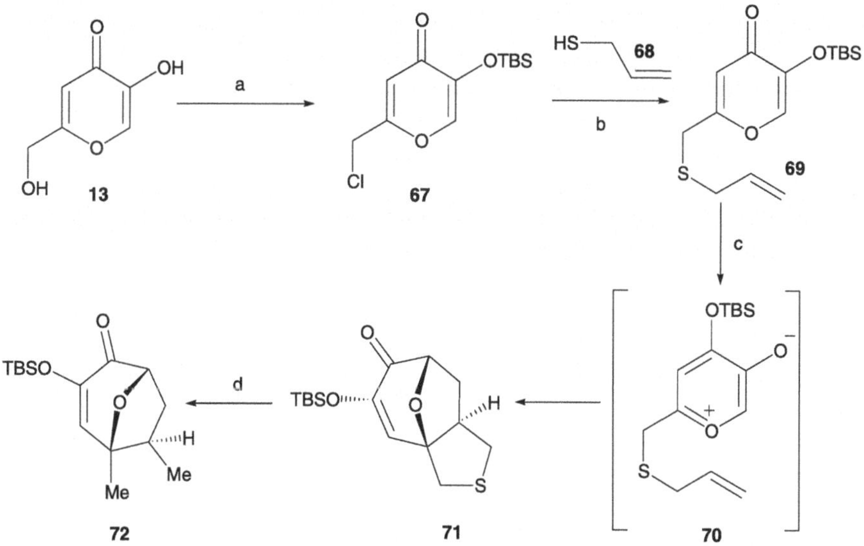

Scheme 19 Reagents and conditions: *a* $SOCl_2$, TBSCl; *b* Et_3N; *c* 145 °C; *d* Raney Ni

when sulfur or silicon tethers were used to connect the alkenes and pyrones (Scheme 19) [24].

Investigations from the same group also showed that 3-(silyloxy)-8-oxabicyclo[3.2.1]oct-3-en-2-ones (e.g., **71**), which are easily assembled by the thermal [5 + 2] cycloaddition between β-(silyloxy)-γ-pyrones and tethered alkenes can be readily transformed into stereochemically rich 2,5-*cis*-disubstituted tetrahydrofurans. The strategy utilizes a two-step process involving a stereoselective addition–alkylation sequence followed by oxidative C – C bond cleavage (Scheme 20) [25].

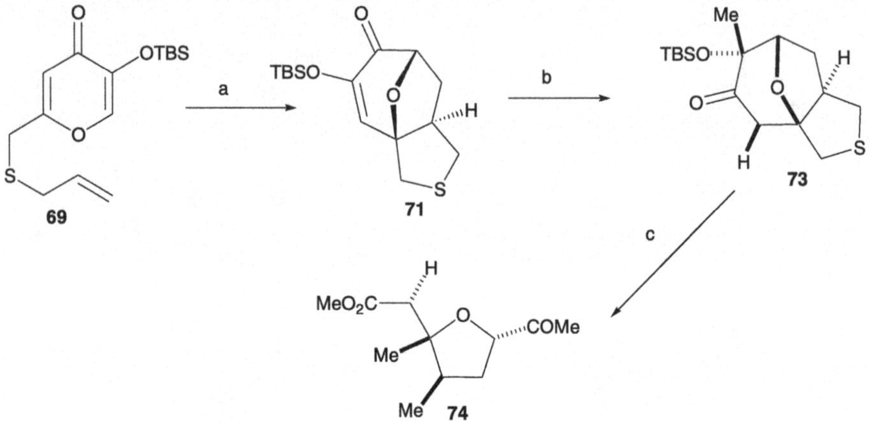

Scheme 20 Reagents and conditions: *a* heat; *b* (i) MeLi; (ii) H^+ or MeI, 87%; *c* (i) Raney Ni (ii) TBAF, $Pb(OAc)_4$, MeOH, 78%

The daphnane diterpenes represent a family of structurally complex oxabridged natural products that collectively exhibit diverse biological activities including neurotrophic, antileukemic, antitumor and insecticidal activity [26]. Wender and coworkers reported a general route to the BC ring system of 12-hydroxy daphnetoxins (e.g., **77**) through an intramolecular [5 + 2] cycloaddition of oxidopyrylium betaines. Here the precursor **76** was prepared from D-ribose **75** and the key cycloaddition was effected by treating **76** with DBU in acetonitrile at 80 °C (Scheme 21) [27].

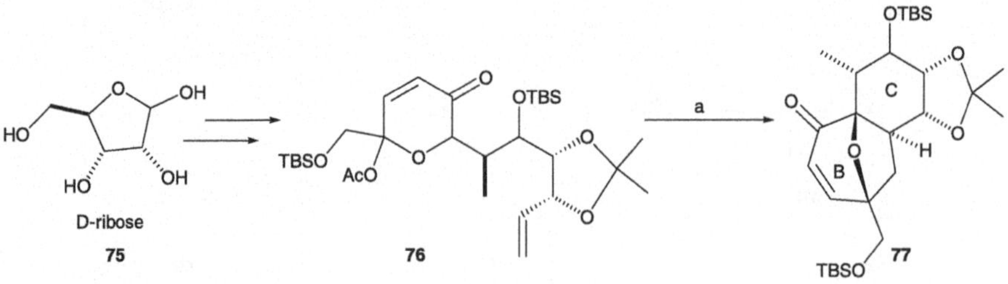

Scheme 21 Reagents and conditions: *a* CH_3CN, DBU, 84%

An efficient synthetic route to the ABC tricyclic core of 1α-alkyldaphnanes **81** has been developed by Wender and coworkers utilizing the intramolecular oxidopyrylium cycloaddition strategy (Scheme 22) [28].

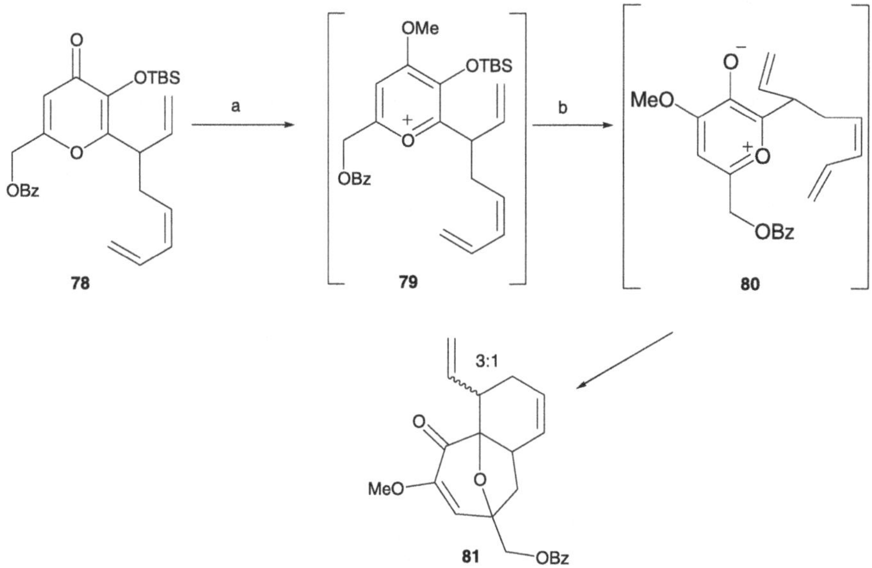

78 79 80 81

Scheme 22 Reagents and conditions: *a* MeOTf, $CDCl_3$ (1 M); *b* CsF, DMF, CH_2Cl_2, 89%

Lee and coworkers have showed that the intramolecular 3-oxidopyrylium betaine-allene cycloaddition proceeded more efficiently than the intermolecular reaction. An example is illustrated in Scheme 23 [21].

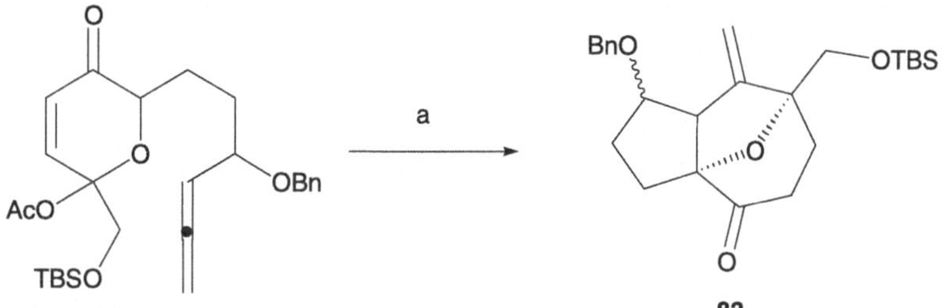

82 83

Scheme 23 Reagents and conditions: *a* DBU, CH_2Cl_2, 81%

D.R. Williams and coworkers have utilized the [3 + 2] intramolecular cycloaddition of a 3-oxidopyrylium ylide toward the synthesis of secodolastanes, and the key step of the reaction is shown in Scheme 24 [29].

D.M. Hodgson et al. have reported a *catalytic* enantioselective tandem oxidopyrylium formation–intramolecular 1,3-dipolar cycloaddition reac-

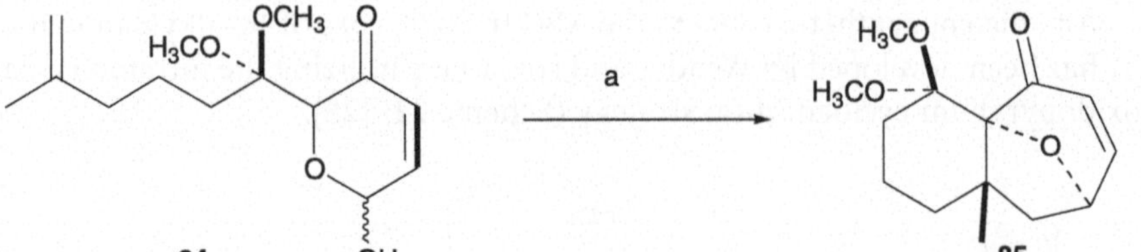

84 85

Scheme 24 Reagents and conditions: *a* (i) AcCl, pyridine, 0 °C, CH_2Cl_2 (ii) DBU, 22 °C, 85%

tion of phthalic anhydride-derived unsaturated diazocarbonyl compounds (Scheme 25) [30].

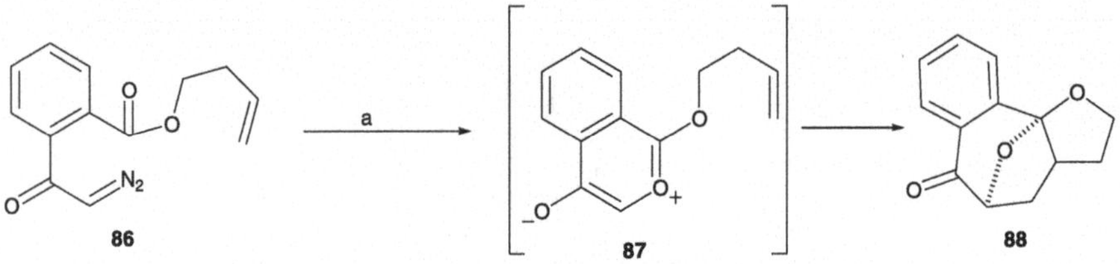

86 87 88

Scheme 25 Reagents and conditions: *a* $Rh_2(OAc)_4$, CH_2Cl_2, 78%

3.3
With 4π Components

As in the case of analogous 3-oxidopyridinium betaines, 4π electron systems such as dienes add across the 2,4-positions of the 3-oxidopyrylium betaine. For example, 3-oxidopyrylium betaine reacts with 2,3-dimethylbutadiene to yield the principal product **90**, which is formed by the addition of diene across the 2,4-position of the ylide along with a small quantity of the 2,6-adduct **91** (Scheme 26) [31].

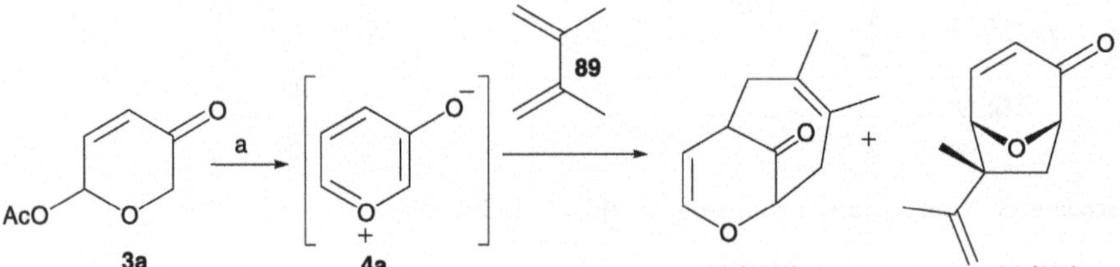

3a 4a 90 (49%) 91 (8%)

Scheme 26 Reagents and conditions: Et_3N, CH_2Cl_2, rt

In the presence of isoprene, 3-oxidopyrylium betaine produced a mixture of regioisomers by the addition across the 2,4- and 2,6-positions (Scheme 27).

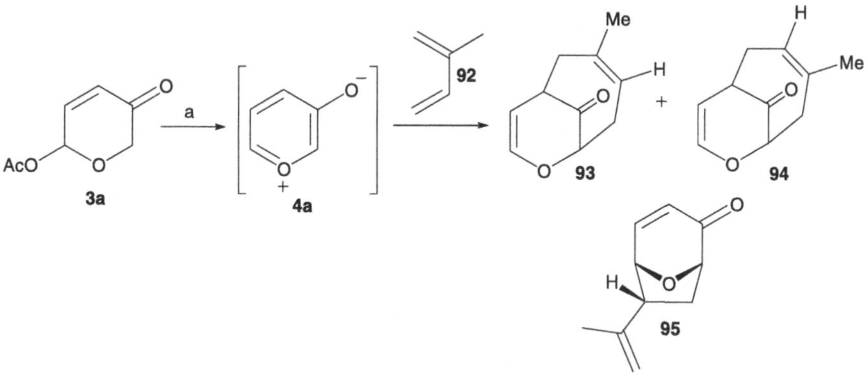

Scheme 27 Reagents and conditions: Et_3N, CH_2Cl_2, rt, 16 h, 62% (total yield)

With cyclopentadiene, cycloaddition across the enone moiety of the pyrylium betaine precursor occurred faster than the formation of the ylide and afforded the product **97** in 75% yield (Scheme 28) [31].

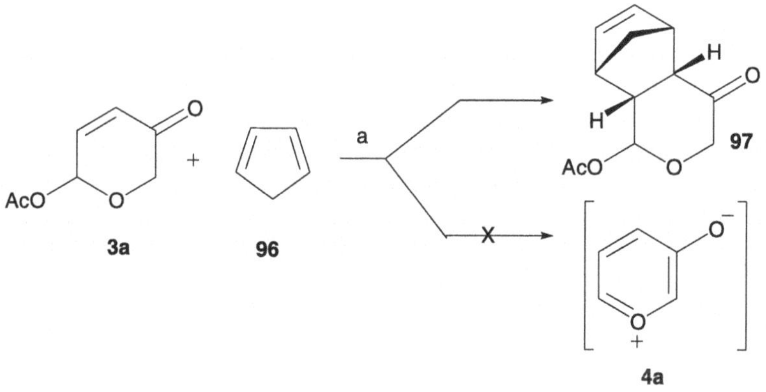

Scheme 28 Reagents and conditions: Et_3N, CH_2Cl_2, rt, 75%

3.4 With 6π Components

K.V. Radhakrishnan et al. have recently reported a facile [6 + 3] cycloaddition of 3-oxidopyrylium betaines with pentafulvenes. This is the first report on the cycloaddition reaction of 3-oxidopyrylium betaine with a 6π component. The reaction of 3-oxidopyrylium betaine generated from pyranulose acetate with 6,6-diphenyl fulvene **1** proceeded through a [6 + 3] cycloaddition mode and afforded the adduct **99**, which then rearranged to the more stable adduct **100** through a 1,5 H-shift (Scheme 29) [32, 33].

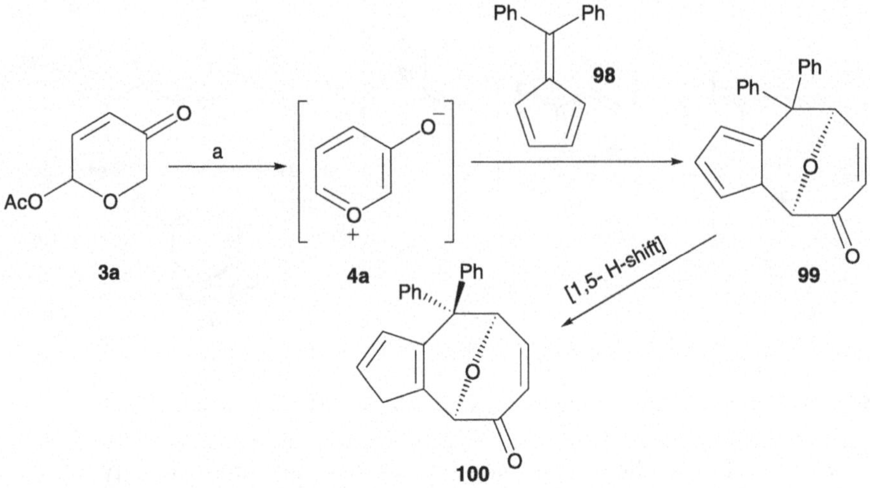

Scheme 29 Reagents and conditions: *a* Et_3N; *b* $CHCl_3$, 50 °C, 6 h, 70%

The reaction was found to be general for a number of pentafulvenes having alkyl, aryl or cyclic substituents at the exocyclic position.

K.V. Radhakrishnan et al. have developed an efficient methodology for the synthesis of novel and versatile 7-5-8 fused oxabridged tricyclic systems through the consecutive dipolar cycloaddition reaction between 3-oxido-pyrylium betaine and pentafulvenes. The reaction of 3-oxidopyrylium betaine with the 5-8 fused oxabridged cyclooctanoid **101**, formed from an initial [6 + 3] cycloaddition, proceeded smoothly, affording a separable mixture of 7-5-8 fused oxabridged tricyclic systems **102** and **103** in 81% yield. The reaction is illustrated in Scheme 30 [34].

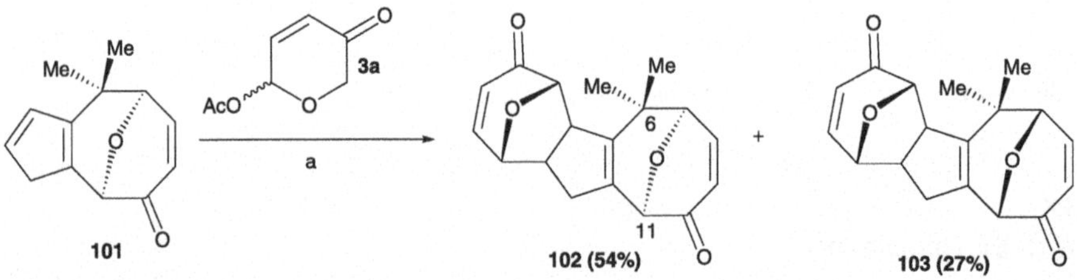

Scheme 30 Reagents and conditions: *a* Et_3N, $CHCl_3$, 50 °C, 6 h, 81%

The work from the above group also uncovered a novel synthesis of 5-8-5 fused tricyclic molecules through the 3-oxidopyrylium betaine–pentafulvene cycloaddition strategy. The dipolar cycloaddition reaction of the azomethine ylide generated from *N*-methoxymethyl-*N*-trimethylsilylbenzylamine **105** in the presence of trifluoroacetic acid, with the [6 + 3] adduct of

3-oxidopyrylium betaine and pentafulvene, proceeded via a [3 + 2] mode and afforded the 5-8-5 fused system **106** in 74% yield (Scheme 31) [35].

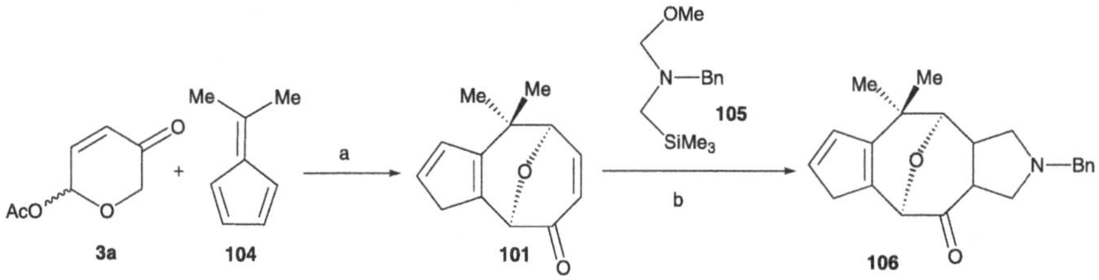

Scheme 31 Reagents and conditions: *a* Et_3N, $CHCl_3$, 50 °C; *b* TFA, CH_2Cl_2, RT, 3 h, 74%

K.V. Radhakrishnan et al. have unraveled a novel and versatile method towards the synthesis of eleven-membered carbocycles through a three-step reaction sequence from 3-oxidopyrylium betaines and pentafulvenes. The [6 + 3] adduct **107** of pentafulvenes with 3-oxidopyrylium betaine on selective reduction, followed by ruthenium catalyzed oxidative cleavage, afforded a novel eleven-membered carbocyclic triketone with a bridging ether linkage (Scheme 32) [36].

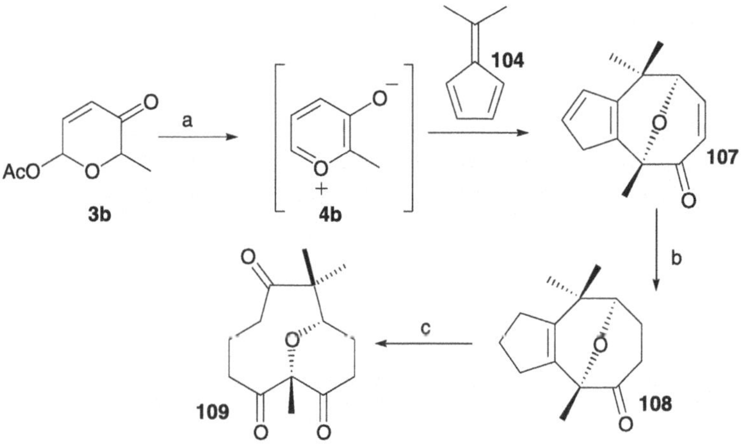

Scheme 32 Reagents and conditions: *a* Et_3N, $CHCl_3$, 50 °C; *b* H_2, Pd/C, EtOAc; *c* $RuCl_3{\cdot}3H_2O$, $NaIO_4$, 74%

4 Applications to Natural Product Synthesis

3-Oxidopyrylium betaine–alkene [5 + 2] cycloaddition provides an interesting and potentially versatile entry into highly functionalized oxabicycles from simple precursors [37]. The synthetic relevance of this reaction, and in particular that of its intramolecular version, are demonstrated by its utility in the synthesis of oxabridged cycloheptanoid fragments of complex natural products [38].

Phorbol is a tigliane diterpene and derivatives possessing the phorbol skeleton hold promise as chemotherapeutic leads due to their antitumor and anti-HIV activity [39]. Wender and coworkers have synthesized a highly flexible precursor **111** of phorbol **112** through an intramolecular pyrylium betaine-mediated cycloaddition and the reaction is depicted in Scheme 33 [40].

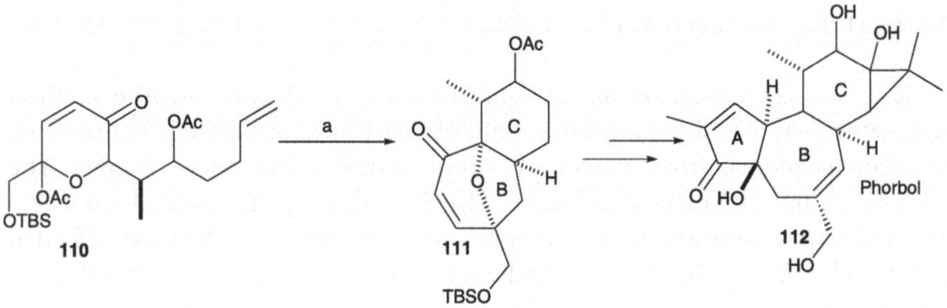

Scheme 33 Reagents and conditions: *a* DBU, CH_3CN

A stereoselective synthesis of the core portion of the diterpene antibiotic guana-castepene **115** was reported by Magnus and coworkers. The highly functionalized 5-8 fused oxabridged framework **114** was generated through an efficient intramolecular cycloaddition of the precursor **113** under reflux conditions in benzene (Scheme 34) [41].

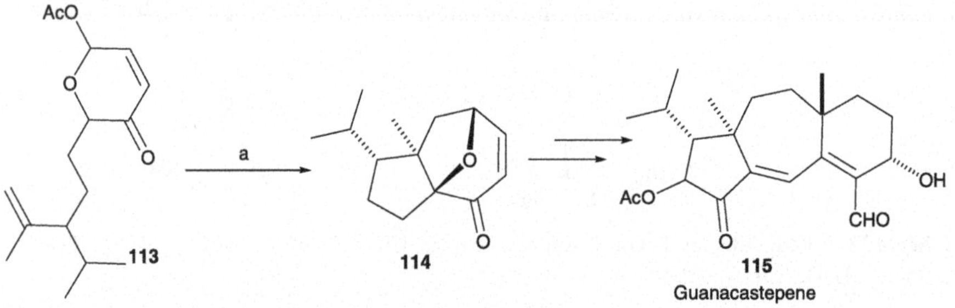

Scheme 34 Reagents and conditions: *a* benzene, reflux, 80%

Using a similar protocol, Magnus and coworkers have reported a stereoselective synthesis of a cyathin diterpene, which has an unusual 5,6,7-ring fused skeleton. Here the pyrylium ylide precursor **118** was synthesized from diethyl adipate and the cyclization was effected by treating with trifluoroacetic acid. The reaction is illustrated in Scheme 35 [42].

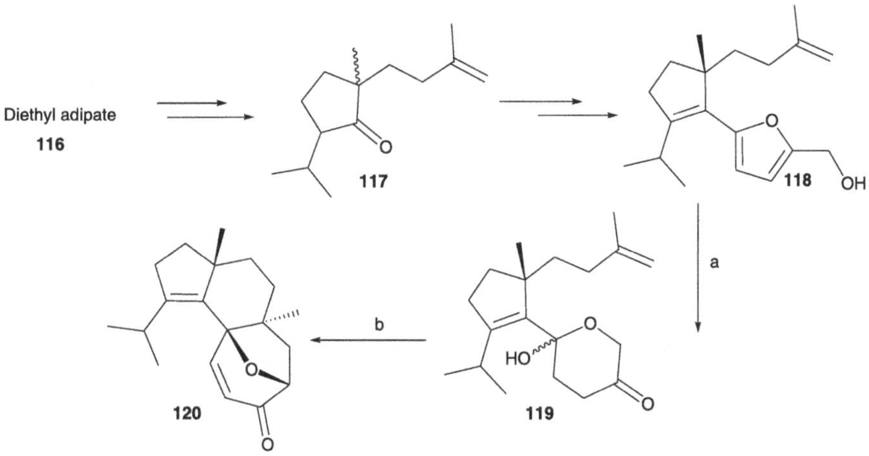

Scheme 35 Reagents and conditions: *a* (i) $O_2/h\nu$, rose bengal (ii) Me_2S; *b* CF_3CO_2H, CH_2Cl_2, 25 °C, 62%

Naoki Ohmori has utilized the intermolecular oxidopyrylium–alkene cycloaddition as one of the key steps in the synthesis of a highly functionalized core structure of phomoidride B (Scheme 36) [43].

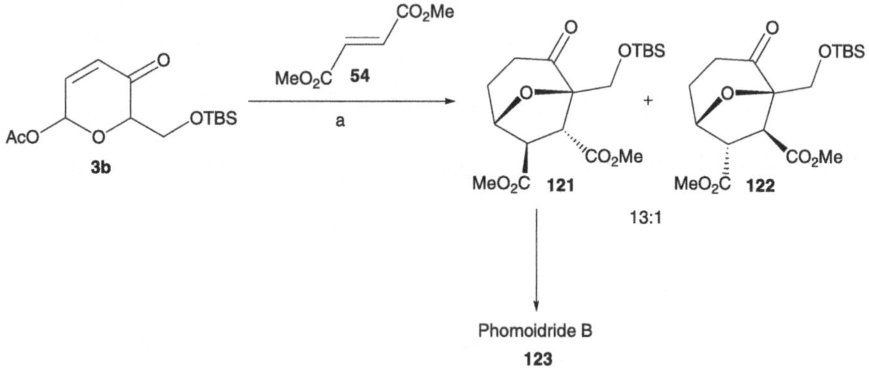

Scheme 36 Reagents and conditions: *a* Et_3N, acetonitrile, reflux, 77%

4.1 Dimerizations of Pyrylium Betaines

Hendrickson and coworkers have observed the formation of a 3-oxidopyrylium betaine dimer via a [5 + 3] cycloaddition (Scheme 37) [9]. The dimer was found to be an eight-membered carbocycle endowed with diverse functionalities around the ring.

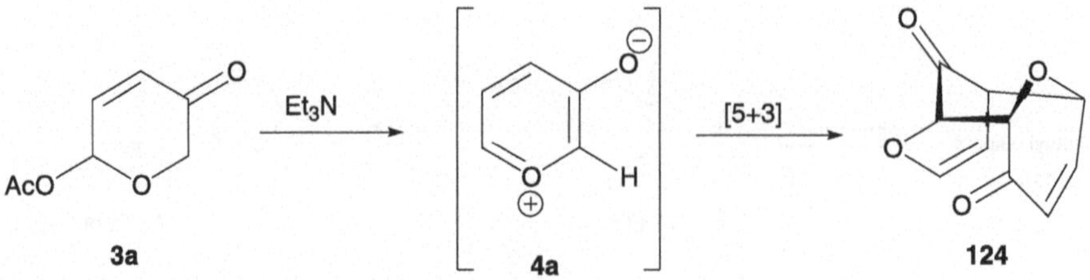

Scheme 37 Reagents and conditions: *a* Et_3N, rt

G.K. Trivedi and coworkers have utilized the self-dimerization of 3-oxido-pyrylium betaines in the stereocontrolled synthesis of highly functionalized cyclooctanoids (Scheme 38) [44].

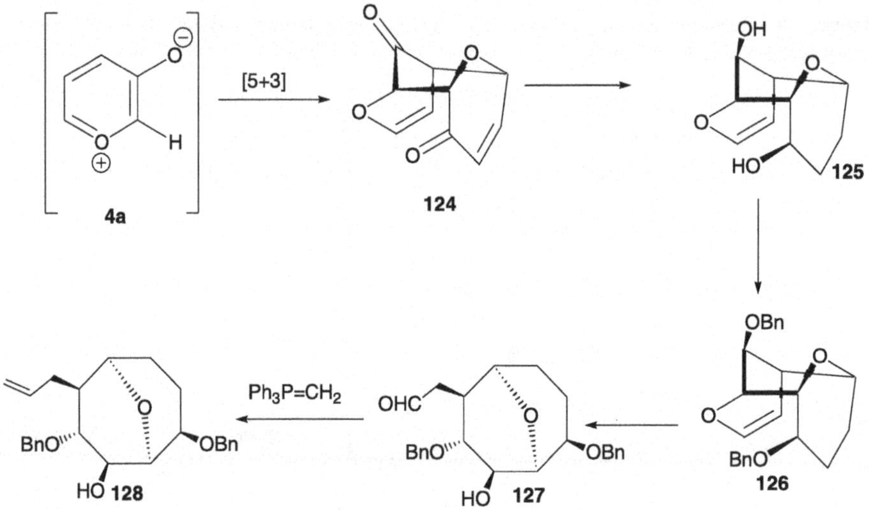

Scheme 38 Reagents and conditions: *a* (i) H_2 – Pd/C, toluene, 90% (ii) $NaBH_4$, MeOH, 10 °C, 68%; *b* NaH, BnBr, Bu_4NI, THF, 77%; *c* Dowex 50W-X4, LiBr, H_2O, CH_3CN, 88%; *d* *n*-BuLi, THF

5 Cycloadditions of Pyridinium Betaines

1,3-Dipolar cycloadditions involving oxidopyridinium betaines have proven to be valuable in organic synthesis. The leading contributions from Katritzky and team-mates have greatly contributed to pyridinium betaine chemistry [45]. They have extensively utilized this cycloaddition strategy for the synthesis of core fragments of tropone alkaloids and related natural products. In addition, dimerizations of suitably substituted pyridinium betaines offered a facile entry into complex heterocyclic frameworks [46]. The 3-oxidopyridiniums are generated from corresponding 3-hydroxypyridinium salts using bases like NaOH, Amberlite IRA-401 (OH^-) resin, or triethylamine (Scheme 39).

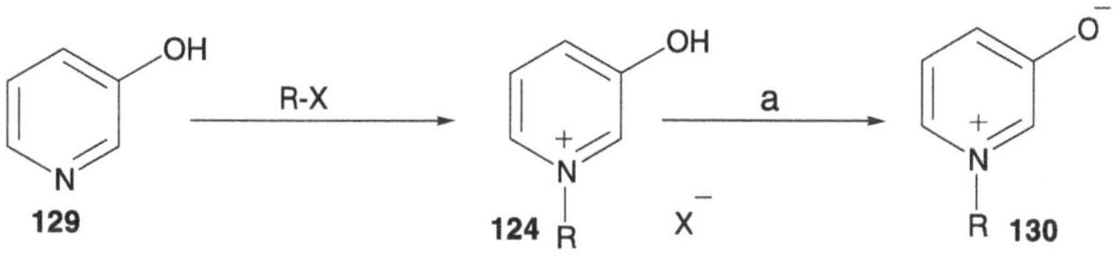

Scheme 39 Reagents and conditions: *a* Et_3N or NaOH

V.K. Aggarwal and coworkers have reported a highly diastereoselective 1,3-dipolar cycloaddition reaction of 3-oxidopyridinium betaines with a C2-symmetric vinyl sulfoxide, *trans*-2-methylene-1,3-dithiolane 1,3-dioxide. High yields coupled with complete diastereoselectivities have been achieved with simple 3-oxidopyrydinium betaines. The regio- as well as diastereoselectivities in these reactions can be improved by using 2-substituted 3-oxidopyridinium betaines (Scheme 40) [15].

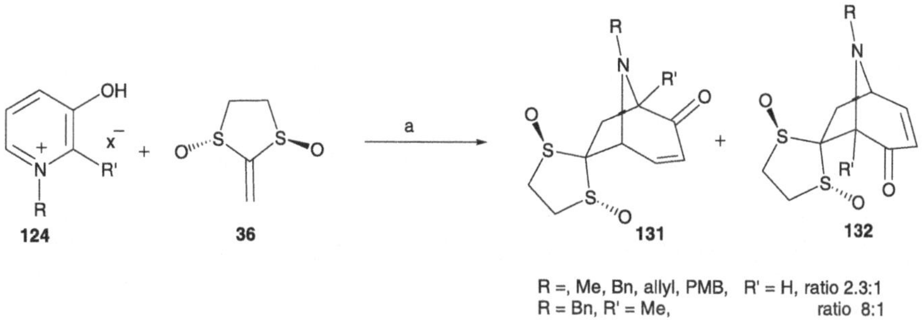

Scheme 40 Reagents and conditions: *a* Et_3N, CH_2Cl_2, rt, 18 h

A.P. Kozikowski et al. have reported an intramolecular dipolar cycloaddition of 3-oxidopyridinium betaine as one of the key steps towards the synthesis of tricyclic cocaine analogs. The reaction is illustrated in Scheme 41 [47].

J.J. Kulagowski et al. have reported asymmetric 1,3-dipolar cycloadditions of homochiral pyridinium betaines towards the synthesis of 8-azabicyclo[3.2.1]oct-3-en-2-ones. The homochiral betaine **138** was prepared from ketone **137** and its cycloaddition with *t*-butyl acrylate was performed in toluene at 95 °C to give, in 70% overall yield, cycloadducts **140–143** with complete regioselectivity (Scheme 42) [48].

Sarains are structurally unusual alkaloids found in marine sponges and were reported to exhibit antitumor, antibacterial, and insecticidal activ-

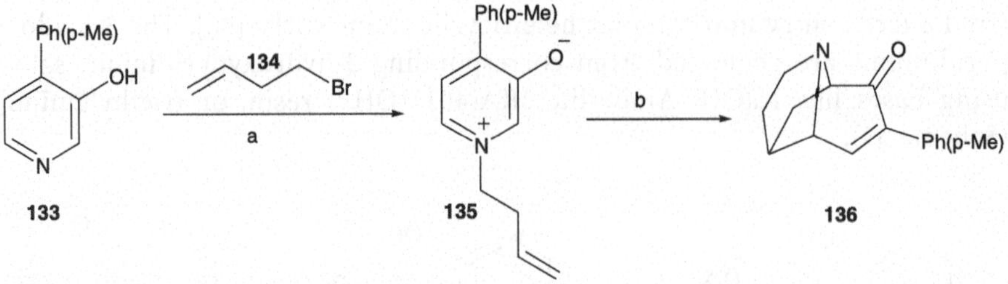

Scheme 41 Reagents and conditions: *a* Amberlite IRA-400 (OH), MeOH, 98%; *b* xylene, reflux, 36%

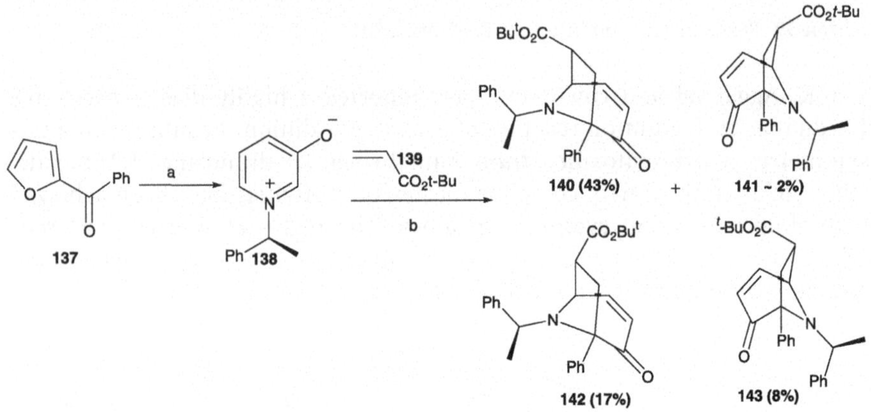

Scheme 42 Reagents and conditions: *a* (i) (*S*)-α-methylbenzylamine, Ti(O – *i*Pr)$_4$, 80 °C then $NaBH_4$, MeOH, rt (95%) (ii) Br_2, THF/H_2O, 0 °C, rt (52%); *b* toluene, 95 °C, 4 days, 70%

ity [49]. J.K. Cha and coworkers have developed a new approach to a suitably functionalized tricyclic core of sarains by means of 1,3-dipolar cycloaddition using 3-oxidopyridinium betaines. The ideal starting material for the sarain precursor **145** was prepared by the dipolar cycloaddition of the pyridinium salt **144** with cyclopentadiene (Scheme 43) [50].

J.L. Mascarenas and coworkers reported a novel approach for the generation of 3-oxidopyridinium betaines. 3-Hydroxy-4-pyridones, which are easily prepared from commercially available 3-hydroxy-4-pyrones, can be readily transformed into 4-methoxy-3-oxidopyridinium ylides by treatment with methyltrifluoromethane sulfonate and subsequent deprotonation with a nonnucleophilic base. These ylides undergo cycloaddition to several electron-deficient alkenes and afford highly functionalized azabicyclo[3.2.1]octane moieties, e.g., **153** (Scheme 44) [51].

M.E. Jung has utilized a highly regioselective 1,3-dipolar cycloaddition of acrylonitrile to *N*-benzyl-3-hydroxypyridinium bromide **154** as the key step towards the total synthesis of the natural product Bao Gong Teng A. The reaction is illustrated in Scheme 45 [52].

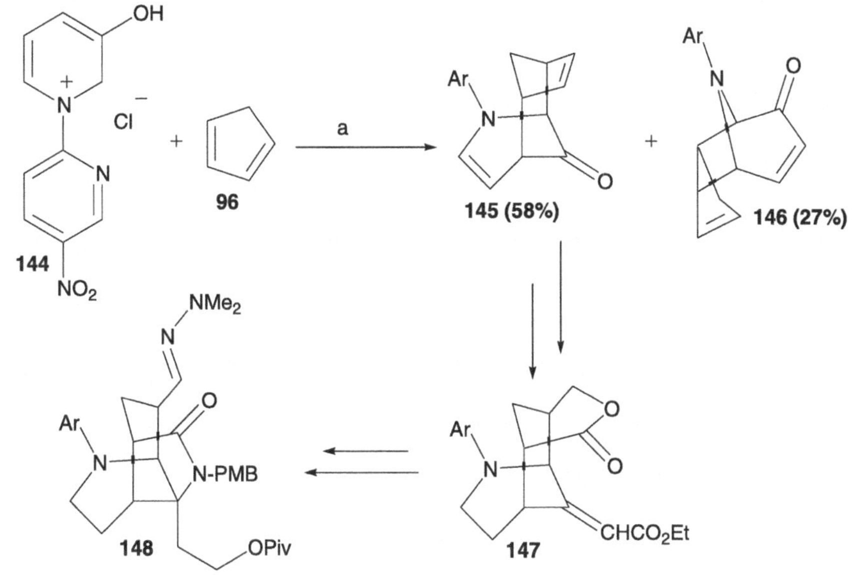

Scheme 43 Reagents and conditions: *a* Et_3N

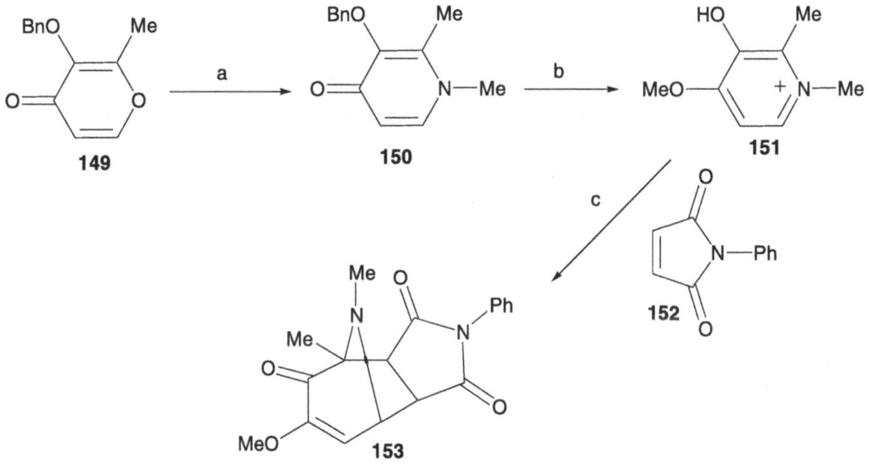

Scheme 44 Reagents and conditions: *a* (i) $MeNH_2$ (ii) hydrogenation, 78%; *b* MeOTf; *c* 2,2,6,6-tetramethylpiperidine (TMP), CH_3CN, reflux, 6 h, 72%

D.Y. Gin and coworkers have reported the generation and dipolar cycloaddition of an analogous 3-oxidopyridinium ylide, oxidoisoquinolinium betaine as the key synthetic steps toward the hetisine alkaloid nominine [53]. When a solution of betaine **159**, prepared from **158**, in THF was heated in a sealed tube at 180 °C, intramolecular cycloaddition to **161** occurred with

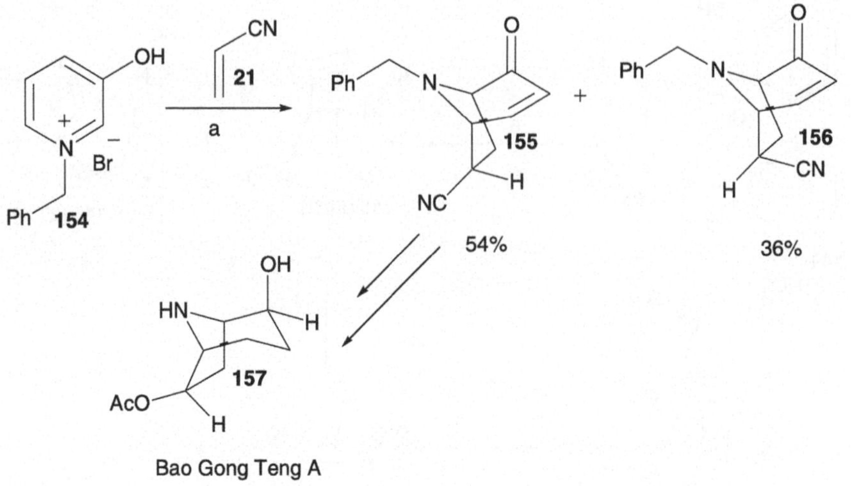

Scheme 45 Reagents and conditions: *a* Et_3N, heat

97% conversion and provided a separable mixture of isomers. This arises due to the differential facial approach of the dipole–dipolarophile partners (Scheme 46) [54].

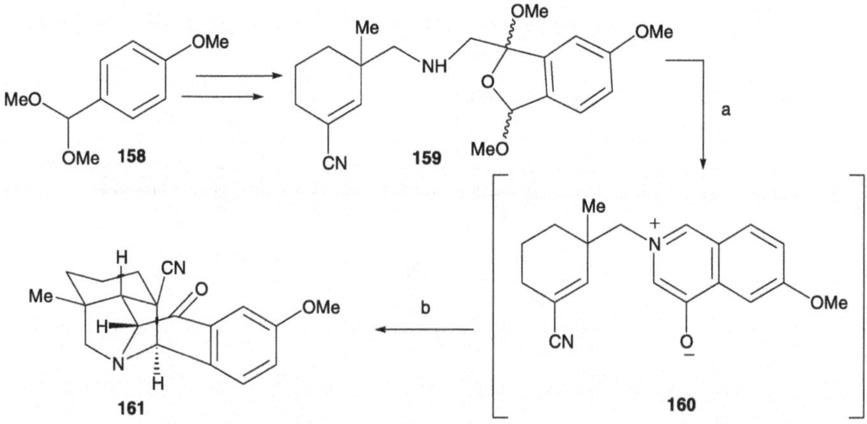

Scheme 46 Reagents and conditions: *a* TFA in CH_2Cl_2, 0 °C, 93%; *b* THF, 180 °C

A novel strategy for the synthesis of the isoschizozygane alkaloid core **166**, based on a 1,4-dipolar cycloaddition reaction of a cross-conjugated betaine **164** was reported by A. Padwa and et al. Here the betaine was generated by the reaction of the thiolactam with carbon suboxide at 25 °C and its cycloaddition was effected by heating up to 120 °C in toluene (Scheme 47) [55].

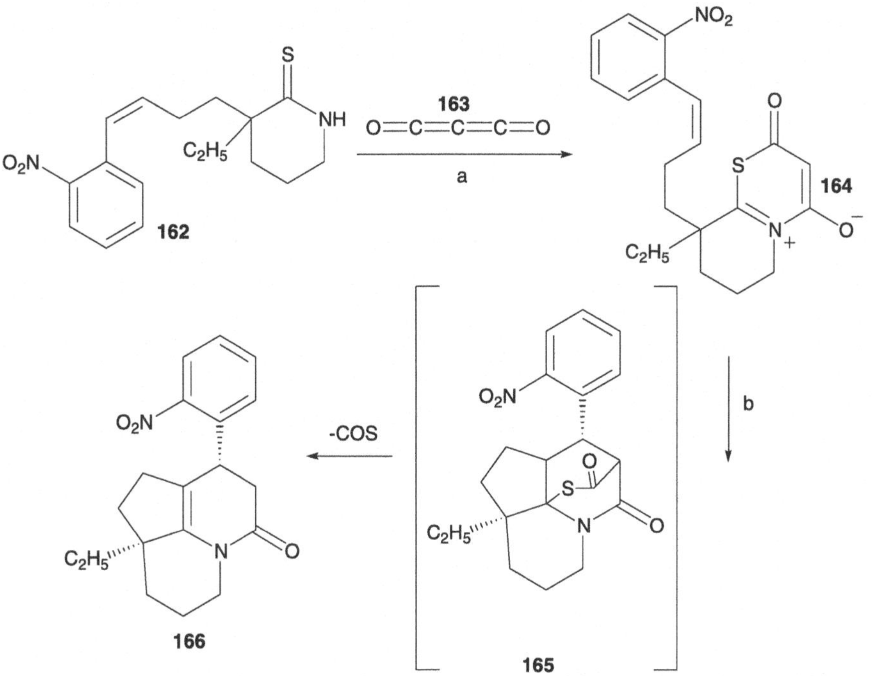

Scheme 47 Reagents and conditions: *a* 25 °C; *b* heat, toluene, 66%

5.1 Dimerizations of Pyridinium Betaines

Unsymmetrical dimers are generally formed spontaneously from 3-oxidopyridiniums when the *N*-substituent is a strongly electron-withdrawing group.

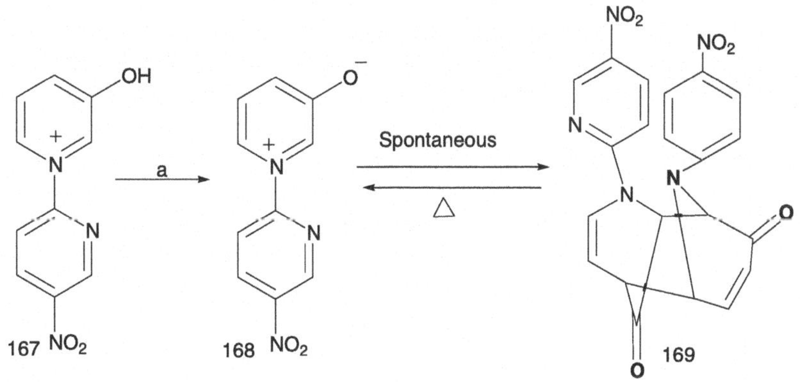

Scheme 48 Reagents and conditions: *a* Et_3N

The dimers are thermally labile and are convenient sources of monomer for further reactions (Scheme 48) [46].

Symmetrical dimmers are formed photochemically from 3-oxidopyridiniums when irradiated above 3500 A^0 (Scheme 49).

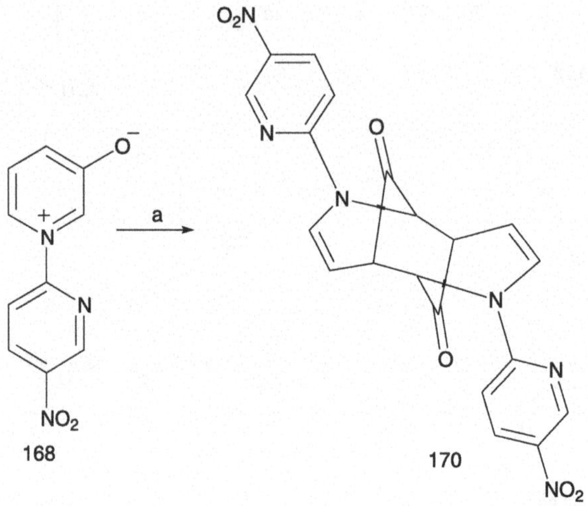

Scheme 49 Reagents and conditions: *a* hν

6
Conclusions

3-Oxidopyrylium betaine–alkene cycloaddition is a simple, well-established method for the rapid assembly of highly functionalized oxabicyclo[3.2.1]octanes from readily available precursors. The oxygen bridge across the seven-membered carbocyclic ring of the adducts imparts conformational rigidity on these otherwise flexible systems. As a consequence, the stereochemistry of further chemical transformations can be controlled. Since the oxygen bridge can be removed easily, the stereoselective synthesis of substituted and biologically important cycloheptanes such as tropones is possible. The rapid generation of molecular complexity in a relatively easy manner has made this approach a highly useful tool in the synthesis of seven-membered rings containing complex natural products. The 1,3-dipolar cycloaddition reactions of 3-oxidopyridnium betaines and olefins have been well studied and utilized for the synthesis of many 8-azabicyclo[3.2.1]octane skeletons, which are the common structural units of biologically active tropane alkaloids.

References

1. Crabb JN, Storr RC (1984) In: Padwa A (ed) 1,3-Dipolar cycloaddition chemistry, vol. 2. Wiley, New York, p 543
2. Waltz AE, Roush WR (1985) Tetrahedron 41:3463
3. Oppolzer W, Petrzilka M (1976) J Am Chem Soc 98:6722
4. Padwa A (1984) 1,3-Dipolar cycloaddition chemistry, vol. I and II. Wiley, New York
5. Katritzky AR, Dennis N (1989) Chem Rev 89:827
6. Hendrickson JB, Farina JS (1980) J Org Chem 45:3359
7. Hendrickson JB, Farina JS (1980) J Org Chem 45:3361
8. Katritzky AR (1989) Chem Rev 89:827
9. Wender PA, Mascarenas JL (1991) J Org Chem 56:6267
10. Sammes PG, Street LJ (1983) J Chem Soc Perkin Trans I 2729
11. Baldwin JE, Mayweg AVW, Pritchard GJ, Adlington RM (2003) Tetrahedron Lett 44:4543
12. Adlington RM, Baldwin JE, Mayweg AVW, Pritchard GJ (2002) Org Lett 4:3009
13. Fishwick CWG, Mitchell G, Pang PFW (2005) Synlett, p 285
14. Marshall KA, Mapp AK, Heathcock CH (1996) J Org Chem 61:9135
15. Aggarwal VK, Grainger RS, Newton GK, Spargo PL, Hobson AD, Adams H (2003) Org Biomol Chem 1:1884
16. Krishna UM, Trivedi GK (2004) Tetrahedron Lett 45:257
17. Snider BB, Grabowski JF (2005) Tetrahedron Lett 46:823
18. Snider BB, Grabowski JF (2006) Tetrahedron 62:5171
19. Wender PA, Mascarenas JL (1992) Tetrahedron Lett 33:2115
20. Ohmori N, Miyazaki T, Kojima S, Ohkata K (2001) Chem Lett, p 906
21. Lee H-Y, Sohn J-H, Kim HY (2001) Tetrahedron Lett 42:1695
22. Delgado A, Castedo L, Mascarenas JL (2002) Org Lett 4:3091
23. Krishna UM, Deodhar KD, Trivedi GK (2004) Tetrahedron 60:4829
24. Rumbo A, Castedo L, Mourino A, Mascarenas JL (1993) J Org Chem 58:5585
25. Mascarenas JL, Rumbo A, Castedo L (1997) J Org Chem 62:8620
26. He W, Cik M, Appendino G, Puyvelde LC, Leysen JE, Dekimpe N (2002) Mini-Rev Med Chem 2:185
27. Wender PA, Bi FC, Buschmann N, Gosselin F, Kan C, Kee J-M, Ohmura H (2006) Org Lett 8:5373
28. Wender PA, D'Angelo N, Elitzin VI, Ernst M, Jackson-Ugueto EE, Kowalski JA, McKendry S, Rehfeuter M, Sun R, Voigtlaender D (2007) Org Lett 9:1829
29. Williams DR, Benbow JW, Allen EE (1990) Tetrahedron Lett 31:6769
30. Hodgson DM, Stupple PA, Johnstone C (2003) Arkivoc, p 49
31. Sammes PG, Street LJ (1983) J Chem Soc Perkin Trans I 1261
32. Radhakrishnan KV, Krishnan KS, Bhadbhade MM, Bhosekar GV (2005) Tetrahedron Lett 46:4785
33. Krishnan KS, Sajisha VS, Anas S, Suresh CH, Bhadbhade MM, Bhosekar GV, Radhakrishnan KV (2006) Tetrahedron 62:5952
34. Krishnan KS, Suresh E, Mathew S, Radhakrishnan KV (2006) Synthesis, p 1811
35. Krishnan KS, Smitha M, Suresh E, Radhakrishnan KV (2006) Tetrahedron 62:12345
36. Krishnan KS, Kuthanapillil JM, Rajan R, Suresh E, Radhakrishnan KV (2007) Eur J Org Chem. doi:10.1002/ejoc.200700767
37. Imafuku K, Arai K (1989) Synthesis, p 501
38. Hong B-C, Wu J-L, Gupta AK, Hallur MS, Liao J-H (2004) Org Lett 6:3453
39. Evans FJ (ed) (1986) Naturally occurring phorbol esters. CRC, Boca Raton, FL

40. Wender PA, Rice KD, Schnute ME (1997) J Am Chem Soc 119:7897
41. Magnus P, Waring MJ, Ollivier C, Lynch V (2001) Tetrahedron Lett 42:4947
42. Magnus P, Shen L (1999) Tetrahedron 55:3553
43. Ohmori N (2001) Chem Commun, p 1552
44. Krishna UM, Deodhar KD, Trivedi GK, Mobin SM (2004) J Org Chem 69:967
45. Katritzky AR, Cutler AT, Dennis N, Sabonji GJ, Rahini-Rastgoo S, Fischer GW, Fletcher IJ (1980) J Chem Soc Perkin Trans I, p 1176
46. Katritzky AR, Boonyarakvanich A, Dennis N (1980) J Chem Soc Perkin Trans I, p 343
47. Smith MP, George C, Kozikowski AP (1998) Tetrahedron Lett 39:197
48. Curtis NR, Ball RG, Kulagowski JJ (2006) Tetrahedron Lett 47:2635
49. Caprioli V, Cimino G, De Guilio A, Madaio A, Scognamiglio G, Trivellone E (1992) Comp Biochem Physiol 103B:293
50. Sung MJ, Lee HI, Chong Y, Cha JK (1999) Org Lett 1:2017
51. Rumbo A, Mourino A, Castedo L, Mascarenas JL (1996) J Org Chem 61:6114
52. Jung ME (1992) J Org Chem 57:3528
53. Peese KM, Gin DY (2006) J Am Chem Soc 128:8734
54. Sakai S, Yamamoto I, Yamaguchi K, Takayama H, Ito M, Okamoto T (1982) Chem Pharm Bull 30:4579
55. Padwa A, Flick AC, Lee HI (2005) Org Lett 7:2925

Top Heterocycl Chem (2008) 13: 99–171
DOI 10.1007/7081_2008_115

Published online: 15 March 2008

Indium Catalyzed Synthesis of Heterocycles via Cycloadditions

J. S. Yadav (✉) · B. V. Subba Reddy · R. Srinivasa Rao · G. Rajendar

Division of Organic Chemistry, Indian Institute of Chemical Technology,
500 007 Hyderabad, India
yadavpub@iict.res.in

Abstract Cycloaddition reactions have found many important synthetic applications in the preparation of biologically important classes of heterocycles. Cycloaddition reactions lead to the formation of new σ-bonds via a cyclic transition state resulting in cyclic products with desired ring size and stereoselectivities. This review collects the synthesis of the most important heterocyclic compounds via cycloadditions under indium catalysis. This review also discusses the mechanistic and stereochemical aspects of these cycloadditions.

Keywords Cycloaddition reactions · Heterocycles · Indium catalysis

Abbreviations

AgOTf	Silver trifluoromethanesulfonate
$AlCl_3$	Aluminum(III) chloride
AZT	Azidothymidine
$BiCl_3$	Bismuth(III) chloride
$Bi(OTf)_3$	Bismuth(III) chloride
$BF_3{\cdot}OEt_2$	Boron trifluoride diethyl etherate
BVDU	Bromovinyldeoxyuridine
$CeCl_3{\cdot}7H_2O$	Cerium(III) chloride heptahydrate
$Ce(OTf)_3$	Cerium trifluoromethanesulfonate
CH_2Cl_2	Dichloromethane
CH_2ClCH_2Cl	1,2-Dichloroethane
CH_3CN	Acetonitrile
$Cu(OTf)_2$	Copper(II) trifluoromethanesulfonate
DDC	Dideoxycytidine
DDI	Dideoxyinosine
DNA	Deoxyribonucleic acid
DQFCOSY	Double quantum filtered correlation spectroscopy
Et_2AlCl	Diethylaluminum chloride
EDA	Ethyl diazoacetate
Et_3N	Triethyl amine
$Gd(OTf)_3$	Gadolinium(III) trifluoromethanesulfonate
Hz	Hertz
HOMO	Highest occupied molecular orbital
$Ho(OTf)_3$	Holmium(III) trifluoromethanesulfonate
HMBC	Heteronuclear multiple bond correlation
HSQC	Heteronuclear single quantum correlation

HCl	Hydrochloric acid
H_2	Hydrogen gas
H_2O	Water
IMDA	Intramolecular Diels–Alder reaction
IMIDA	Intramolecular imino-Diels–Alder reaction
$InBr_3$	Indium(III) bromide
$In(O^tBu)_3$	Indium(III) *tert*-butoxide
$InCl_3$	Indium(III) chloride
InF_3	Indium(III) fluoride
InI_3	Indium(III) iodide
$In(OTf)_3$	Indium(III) trifluoromethanesulfonate
LDA	Lithium diisopropylamide
LUMO	Lowest unoccupied molecular orbital
MeOH	Methanol
$MgBr_2$	Magnesium(II) bromide
$MgSO_4$	Magnesium(II) sulphate
MS	Molecular sieves
MW	Microwave
NMR	Nuclear magnetic resonance
nOe	Nuclear Overhauser effect
NOESY	Nuclear Overhauser effect spectroscopy
PANI-In	Polyaniline supported indium(III) trichloride
Pd/C	Palladium on carbon
^iPrOH	*Iso*-propanol
RNA	Ribonucleic acid
rt	Room temperature
$Sc(OTf)_3$	Scandium(III) trifluoromethanesulfonate
$Sm(OTf)_3$	Samarium(III) trifluoromethanesulfonate
$SnCl_4$	Tin(IV) chloride
$TiCl_4$	Titanium(IV) chloride
$TMSN_3$	Trimethylsilyl azide
TMSOTf	Trimethylsilyl trifluoromethanesulfonate
$YbCl_3$	Ytterbium(III) chloride
$Yb(OTf)_3$	Ytterbium(III) trifluoromethanesulfonate
YCl_3	Yttrium(III) chloride
$ZnBr_2$	Zinc(II) bromide
$ZnCl_2$	Zinc(II) chloride
ZnI_2	Zinc(II) iodide
$Zn(OTf)_2$	Zinc(II) trifluoromethanesulfonate

1 Introduction

Cycloaddition reactions are among the most important and useful processes in organic chemistry as they enable the construction of complex polycyclic compounds, often with a high degree of regio- and stereocontrol [1]. In general cycloaddition reactions are the essence of organic synthesis for carbon–carbon and carbon–heteroatom bond formation, which also provide the foundation for generating more complicated organic compounds from simpler ones. According to nomenclature, organic cycloaddition reactions are differentiated as electronic and topological in nature [2]. Electronic notation represents the number of π-electrons of two fragments involved in a reaction, usually represented as $[\pi m + \pi n]$ and topological notation represents the number of atoms of two fragments involved in addition and usually represented as $[m + n]$ (Table 1).

But for the convenience, notation by topology is mainly preferred for two reasons. Firstly, there are no mechanistic implications, i.e., whether it is a concerted or stepwise reaction, and secondly, it is also readily applicable to the so-called cheleotropic and multi-component cycloadditions [3, 4]. The mechanism for these cycloadditions follows either a concerted or stepwise pathway. The first is most commonly used in the traditional concerted reactions such as the Diels–Alder reaction and the 1,3-dipolar cycloaddition, which are electronically equivalent.

A majority of the compounds produced by nature have heterocyclic rings as part of their structures and are found as key components in biological systems. These heterocycles are the basis of many essential pharmaceuticals, and of many physiologically active natural products. These aspects forced many synthetic chemists to enter the field and to develop new synthetic methodologies for the construction of heterocyclic compounds with control of relative and absolute configuration. Among a variety of new synthetic methods, metal-catalyzed cycloaddition reactions are an important tool for the generation of heterocyclic compounds, since these can directly construct complicated molecules from readily accessible starting materials under mild conditions with desired ring size and stereoselectivities.

Table 1 Characterization of cycloaddition reactions by electronic and topological indicators

	Types of characterization Electronic (Woodward–Hoffmann)	Topological (Huigen)	Cycloadduct
Diels–Alder reaction	$[\pi 4+\pi 2]$	[4+2]	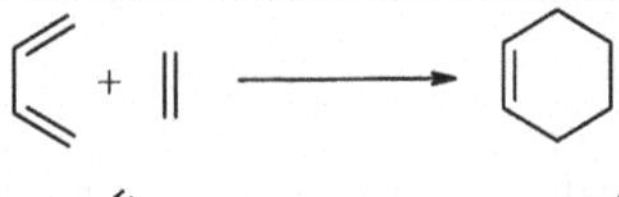
1,3-Dipolar addition	$[\pi 4+\pi 2]$	[3+2]	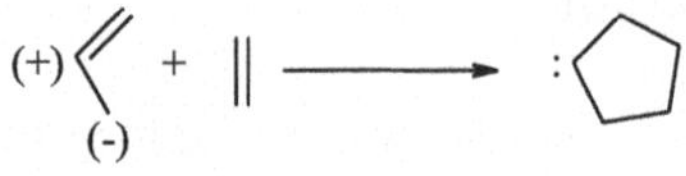
Chelotropic reaction	-	[m+1] $m = 2, 4, 6, \ldots$	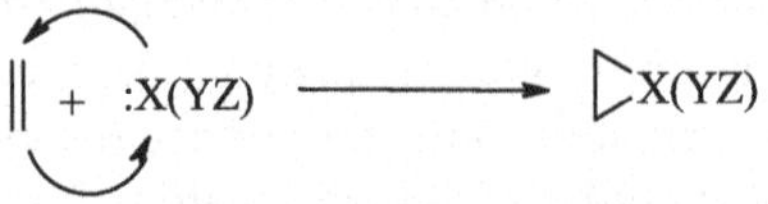

Lewis acids are often employed in the reactions. Indium and its salts have advantages when compared to their aluminum and boron counterparts which are often employed in these reactions in more than stoichiometric amounts. Moreover, these traditional Lewis acids are moisture sensitive and easily decomposed or deactivated in the presence of even a small amount of water, hence cannot be recovered and reused. Furthermore, the residues of these Lewis acids after the reaction may induce some serious environmental problems. In contrast, indium salts are stable in water and are reusable. Among the indium salts $InCl_3$, $InBr_3$, $In(OTf)_3$ are widely used in organic synthesis for various functional group transformations [5–14]. Indium salts also catalyze the synthesis of different heterocycles via cyclocondensation reactions and cyclizations [15–23]. This review only highlights recent developments in the application of indium and its salts as catalysts in the synthesis of heterocyclic compounds by cycloaddition reactions. These indium catalysts provide new opportunities for highly selective cycloaddition reactions, since complexation of the indium to the diene, or dienophile significantly modifies the reactivity of the moiety, opening the way for improved reactivity and novel chemistry.

The applications of cycloadditions in organic synthesis have been the subject of several review articles [24–28] where the fundamentals of these reactions have been described.

2 Three-Membered Rings

2.1 Aziridine-2-carboxylates

The aziridine ring is a versatile building block for organic synthesis, which provides a convenient entry to the stereoselective preparation of many biologically active functionalized amino compounds such as amino acids, β-lactam antibiotics, and alkaloids [29–31]. *N*-Arylidene anilines **1** reacted with ethyl α-diazoacetate (EDA) **2** in the presence of 2 mol% of $InCl_3$ in CH_2Cl_2 at

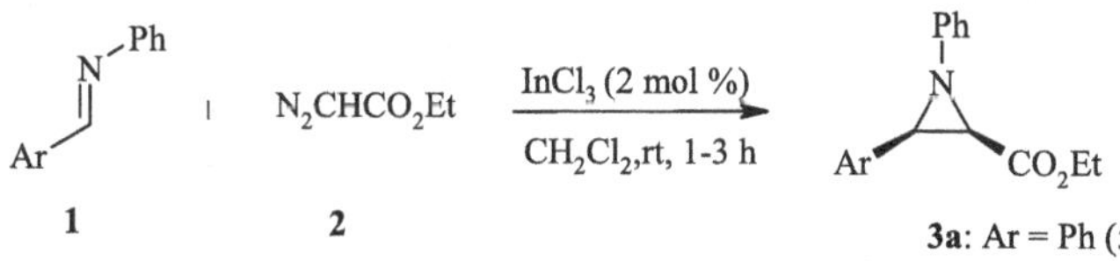

3a: Ar = Ph (50%, *cis* only)
3b: = *p*-Cl-Ph (48 % *cis* only)
3c: = *p*-Me-Ph (45% *cis* only)
3d: = *p*-NO_2-Ph (65% *cis/trans* 80/20)

Scheme 1 Synthesis of *cis*-aziridine carboxylates

room temperature for 1–3 h and produced the corresponding *cis*-aziridine carboxylates **3a–d** together with the corresponding enamine by-products (Scheme 1).

All the aziridines produced were either exclusively as the *cis* isomer **3a–c** or with a high *cis/trans* ratio **3d** [32]. Figure 1 illustrates the mechanism consistent with the above results. Activation of the imine by complexation with the Lewis acid followed by nucleophilic addition of EDA results in formation of intermediate **6** [33, 34]. Subsequent ring closure and loss of N_2 provides aziridines **3**. By-products **4** and **5** result from the 1,2-migration of either the aryl or H-substituent to the incipient carbocation to yield initially **4′** and **5′**. For aromatic imines, the ratio of aziridines to migration products is strongly dependent on the nature of the aryl ring. Thus imines derived from electron-deficient or weakly electron-rich aldehydes gave the best results in keeping with the usual migratory aptitudes for these substituted aryl groups, whereas those derived from strongly electron-rich aldehydes, e.g.,

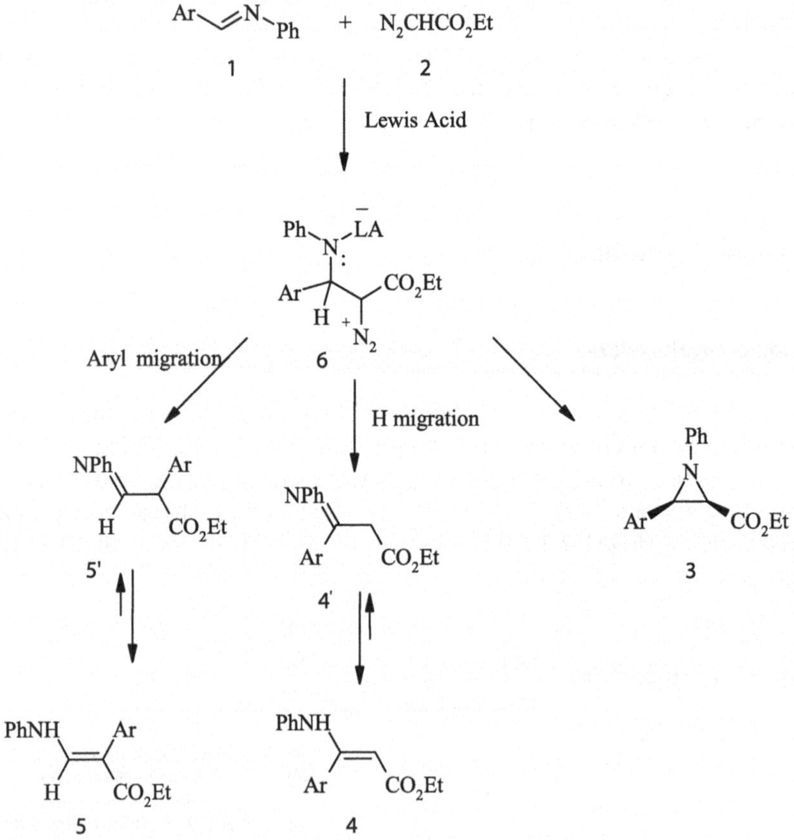

Fig. 1 Mechanism for the formation of *cis*-aziridine carboxylates

p-anisaldehyde, gave complex product mixtures having large amounts of the enamine side products **4** and **5**. In another limitation, aldimines prepared from electron-rich anilines, e.g., *p*-anisidine, when reacted with **2** in the presence of 2 mol % $InCl_3$ led to the darkening of the reaction mixture with no aziridine formation.

2.2 Aziridine-2-phosphonates

Aziridine-2-phosphonates are useful intermediates in the synthesis of biologically important molecules such as α-aminophosphonates, utilized in medicine and agriculture as enzyme inhibitors, antibiotics, clonal antibodies, pesticides and herbicides [35–38].

Diastereomeric aziridine-2-phosphonates **8a–i** (*cis*) and **9a–i** (*trans*) have been prepared by aziridination of substituted arylimines **1** with diisopropyl diazomethylphosphonate **7** in the presence of 10% $In(OTf)_3$ at 0 °C favoring *cis* isomers in overall good yields (Scheme 2) [39], and the results were summarized in Table 2. Among $Zn(OTf)_2$, BF_3cdotOEt_2, $AlCl_3$, $TiCl_4$, $Cu(OTf)_2$, $In(OTf)_3$ screened for this reaction, $In(OTf)_3$ showed better results in terms of yields and stereoselectivity.

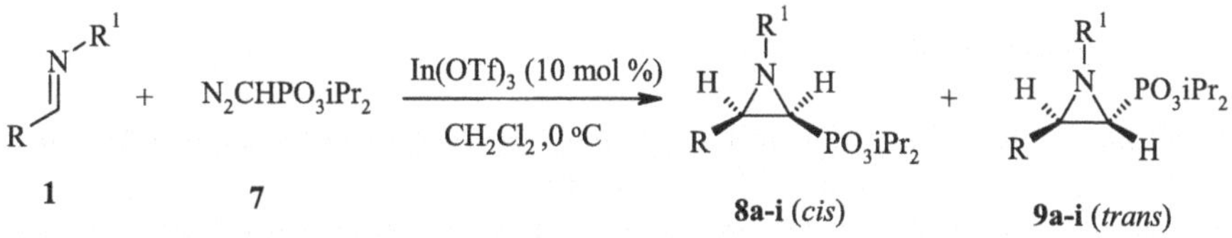

Scheme 2 Synthesis of aziridine phosphonates

Table 2 $In(OTf)_3$ catalyzed synthesis of aziridine phosphonates

Entry	Aryl imines (1) R	R^1	Product ratio **8:9** *cis:trans*	Overall yield % (**8+9**)
a	C_6H_5	C_6H_5	*cis* only	40
b	C_6H_5	4-OMe–C_6H_4	1:1	42
c	C_6H_5	4-CF_3–C_6H_4	–	40
d	C_6H_5	$CHPh_2$	13:1	83
e	4-Me–C_6H_4	$CHPh_2$	4:1	56
f	4-OAc–C_6H_4	$CHPh_2$	4:1	86
g	4-Cl–C_6H_4	$CHPh_2$	6:1	91
h	4-CF_3–C_6H_4	$CHPh_2$	2:1	82
i	4-NO_2–C_6H_4	$CHPh_2$	2:1	80

These (*N*-benzhydryl)-aziridine-2-phosphonates **8** and **9** have been used for the preparation of [1-amino-2-(4-hydroxy-phenyl)-ethyl]-phosphonic acid hydrochloride, a phosphonic acid analog of the tyrosine **11** [40–43]. For example hydrogenolytic cleavage of aziridine **8f** was achieved in methanol under hydrogen flow in the presence of 10% palladium on carbon at room temperature to afford the corresponding diisopropyl α-aminophosphonate **10** in 80% yield, and subsequent hydrolysis of **10** by treatment with 6 N HCl at 90 °C afforded [1-amino-2-(4-hydroxy-phenyl)-ethyl]-phosphonic acid hydrochloride **11** in 87% yield (Scheme 3).

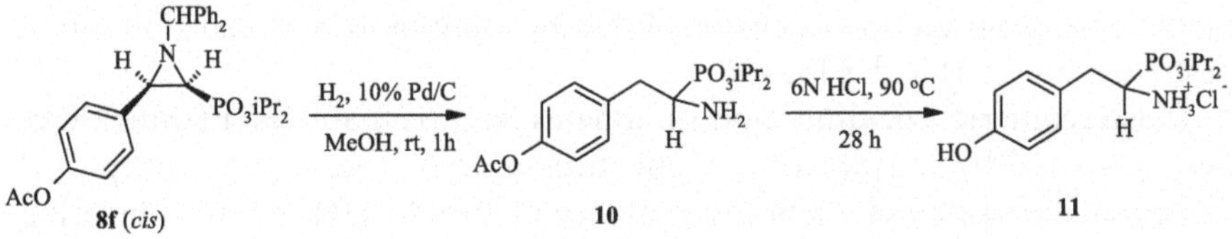

Scheme 3 Synthesis [1-amino-2-(4-hydroxy-phenyl)-ethyl]-phosphonic acid hydrochloride 11

3
Four-Membered Rings

Alkaloids such as sceptrin and nigramide R isolated from terrestrial and marine species possess a cyclobutane moiety (Fig. 2). These compounds exhibit antimicrobial, antibacterial, anticancer, antinociceptive, antidepressant, antiinflammatory, hepatoprotective, and intestinal permeability properties [44].

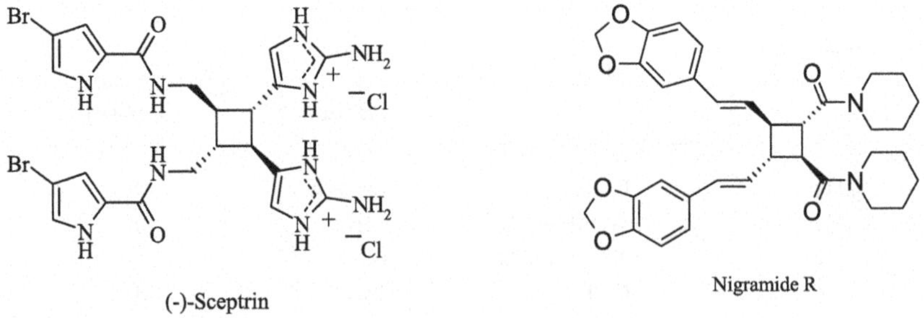

Fig. 2 A few biologically active cyclobutane alkaloids

Owing to the important biological activities of these compounds, the synthesis of *trans*-cyclobutane derivatives **14a–c** via [2+2] cycloaddition of

Table 3 $InCl_3$ catalyzed synthesis of *trans*-cyclobutane derivatives

Alkenes (12a–c) R	R^1	Methyl vinyl ketone **13**	Time (h)	Product	Yield (%)
H	OMe	13	5.5	**14a**	79
OMe	OMe	13	4.0	**14b**	82
OMe	OMe	13	3.5	**14c**	85

methyl vinyl ketone **13** with different electron-rich olefins **12a–c** in the presence of 10 mol % $InCl_3$ in high yields (79–85%) was developed and the results were summarized in Table 3 (Scheme 4) [45].

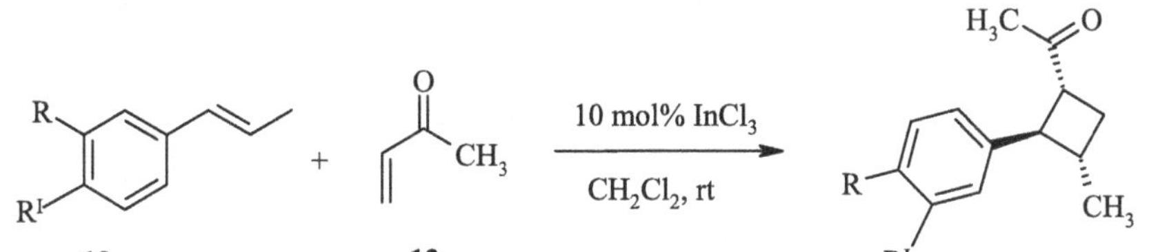

12a-c **13** **14a-c**

Scheme 4 $InCl_3$ catalyzed synthesis of *trans*-cyclobutane derivatives

4 Five-Membered Rings

4.1 1,3-Dipolar Cycloadditions

The 1,3-dipolar cycloaddition reaction has been extensively utilized for the synthesis of five-membered heterocyclic rings [46, 47]. The greatest advantage of this methodology is the construction of polyatomic heterocyclic five-membered rings, which are main precursors of biologically important compounds [48]. The history of 1,3-dipoles goes back to 1883 when Curtius discovered diazoacetic ester [49]. Five years later his younger colleague Buchner studied the reaction of diazoacetic ester with α,β-unsaturated esters and described the first 1,3-dipolar cycloaddition reaction [50]. It was demonstrated for both Diels–Alder and 1,3-dipolar cycloaddition reactions that a four-electron component reacts with a two-electron component, i.e., $[\pi 4 + \pi 2]$. But in the first case the four electrons are distributed among four atoms of the diene, which gives rise to a six-membered cycloadduct, whereas in the second case the four electrons are part of a three-atom fragment, the

1,3-dipole, and a five-membered ring is formed. Different kinds of 1,3-dipoles are defined as an X–Y–Z structure that undergoes 1,3-dipolar cycloaddition reactions and is portrayed by a dipolar structure as outlined in Fig. 3 [51–53]. A challenge of 1,3-dipolar additions is control of regio-, diastereo-, and enantioselectivity. The stereochemistry of the 1,3-dipolar cycloaddition reaction can be controlled either by choosing the appropriate substrates or controlling the reaction by a metal complex acting as a catalyst [54].

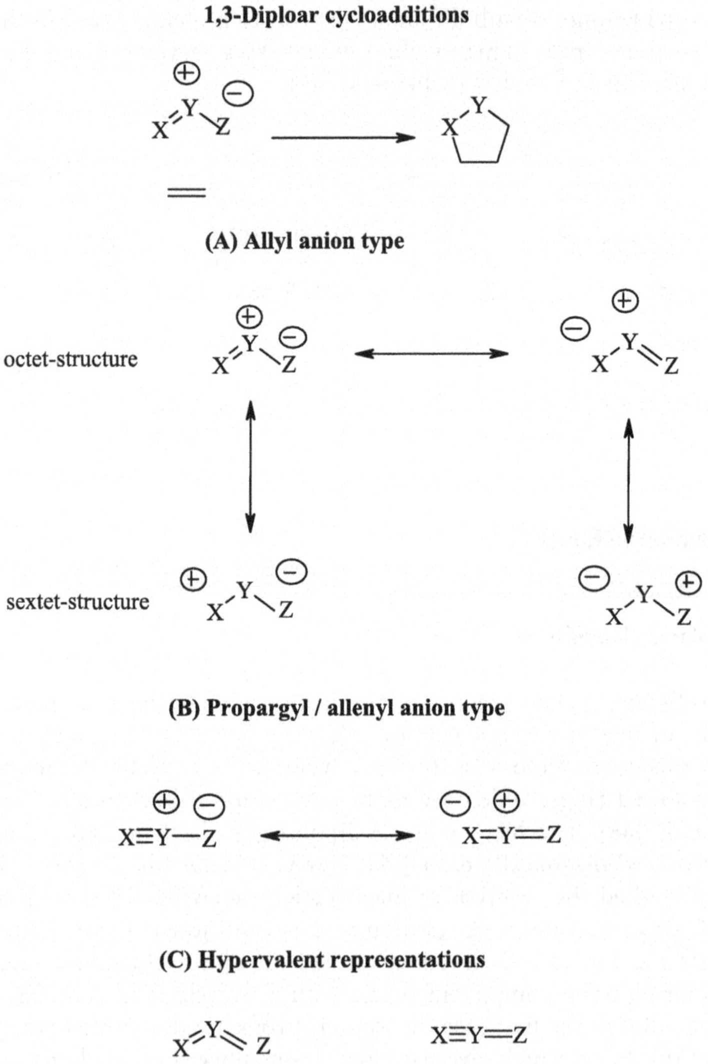

Fig. 3 The basic resonance structures of 1,3-dipoles

4.1.1 Dihydrobenzofurans

Neolignans are a group of secondary metabolites, structurally characterized by the presence of two aryl propanoid units [55–57]. One of these groups possesses the dihydrobenzofuran skeleton with *trans* stereochemistry. Many of these neolignans have been used as potent lead compounds for the development of antiangiogenic, antileishmanial, antiprotozoal, antirheumatic, antifungal, antitumor, antiviral (including HIV), antiinflammatory and hypolipidemic agents [55]. Kadsurenone, 4-*O*-methylcedrusin, conocarpan, sisymbrifolin, Woorenoside IV, 8-*O*-4-dehydro-diferulic acid [58–63] are a few important biologically active 2,3-dihydrobenzofurans containing natural neolignans (Fig. 4).

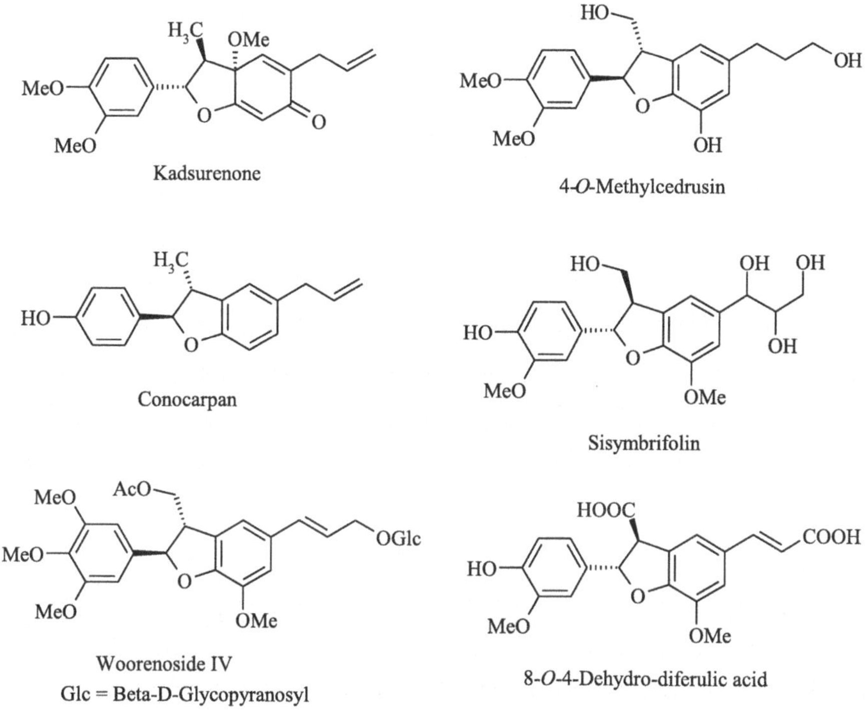

Fig. 4 A few biologically important natural neolignans

So far various multi-step syntheses have been developed towards the synthesis of neolignans [56]. Hence, one-pot [3+2] cycloadditions of various electron-rich olefins (**12a–d**) with substituted *p*-quinones (**15a–b**) catalyzed by $InCl_3$ (10 mol %) have been developed for the synthesis of *trans*-2,3-dihydrobenzofuran derivatives (**16a–f**) at room temperature (Scheme 5) [45].

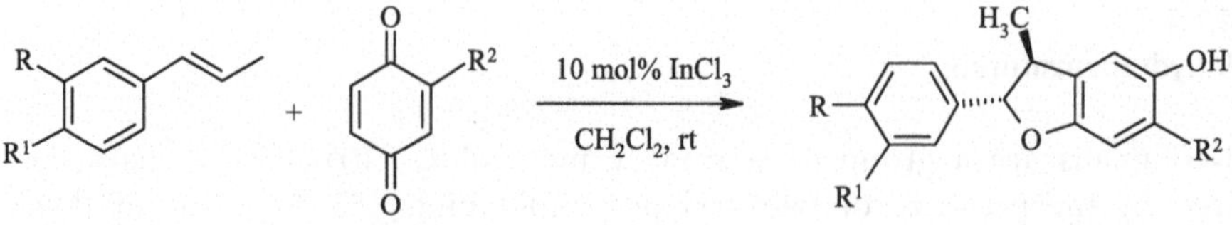

Scheme 5 $InCl_3$ catalyzed synthesis of *trans*-dihydrobenzofurans

The scope and generality of this method is presented in Table 4. High *trans* selectivity, mild reaction conditions, simplicity in operation, high yields (89–94%) and reaction rates, and clean reaction profiles make this method a useful and attractive process for the synthesis of biologically important neolignan derivatives incorporating the 2-aryl-2,3-dihydrobenzofuran framework.

Table 4 $InCl_3$ catalyzed synthesis of *trans*-dihydrobenzofurans

Alkenes (**12a–d**) R	R^1	Quinones (**15a–b**) R^2	Time (h)	Product (**16**)	Yield (%)
H	OMe	H	2.5	**16a**	89
OMe	OH	H	4.0	**16b**	85
OMe	OMe	H	3.5	**16c**	92
–OCH₂O–		H	3.0	**16d**	94
H	OMe	Me	5.0	**16e**	87
OMe	OH	Me	5.5	**16f**	84

4.1.2 Pyrazoles

Pyrazoles are known not only as potent insecticides and herbicides, but also as antitumor, antiinflammatory, antimicrobial and antipsychotic agents [64–67]. Pyrazole derivatives **19a–i** (major) and **20a–i** (minor) have been prepared for the first time by intermolecular 1,3-dipolar cycloaddition or [3+2] cycloaddition of α-diazocarbonyl compounds **17a–f** with alkynes **18a–d** bearing a carbonyl group at the neighboring position in the presence of 20 mol % of $InCl_3$ catalyst in water. Alkyl diazoacetates exclusively afforded thermodynamically more stable pyrazoles **19a–i**, whereas aryldiazoacetates afforded as a mixture of pyrazoles with **19** as the major and **20** as minor products. This reaction did not yield any products with phenyl acetylene and 1,3-dicarbonyl diazo compounds; the results are presented in Table 5 (Scheme 6) [68].

Table 5 $InCl_3$ catalyzed synthesis of pyrazoles

Entry	R^1	R^2	R^3	Product ratio **19:20**	Overall yield (%) (19+20)
a	Ph	OMe	OMe	91:9	89
b	4-OCH_3-C_6H_4	OMe	OMe	94:6	94
c	4-F-C_6H_4	OMe	OMe	89:11	90
d	3-CF_3-C_6H_4	OMe	OMe	86:14	92
e	H	OEt	OEt	100:0	87
f	H	OEt	Me	100:0	81
g	H	OEt	OMe	100:0	79
h	H	Me	OMe	100:0	43
i	CH_3CO	Me	OMe	–	–

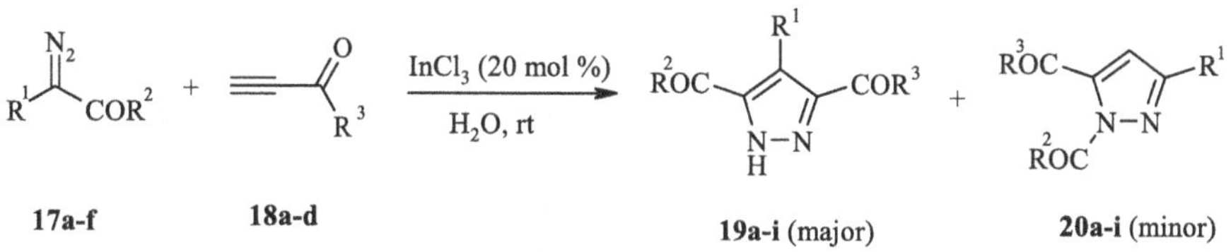

Scheme 6 Synthesis of pyrazoles

The possible mechanism for the formation of the products **19** and **20** has been tentatively illustrated in Scheme 7. The formation of the minor product **20** proceeds by an initial 1,3-dipolar cycloaddition followed by a subsequent 1,3(5)-carboxylate shift of the initially formed 3,5-dicarboxylate-3-aryl-3*H*-pyrazole. The formation of the major product **19** could proceed by three steps: (1) an initial 1,3-dipolar cycloaddition, (2) a subsequent 1,5-aryl shift of initially formed 3,5-dicarboxylate-3-aryl-3*H*-pyrazole to give the intermediate

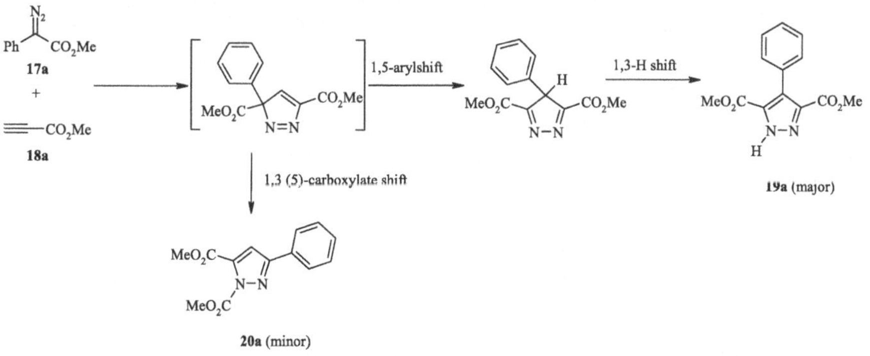

Scheme 7 The possible mechanism for the formation pyrazoles

4*H*-pyrazole, and (3) a 1,3-hydrogen shift of the 4*H*-pyrazole intermediate. The aryl migration product **19** predominates in all cases, and a small electronic effect of the substituent on the aryl group was observed. An aryl group with electron-donating substituents has a higher migratory aptitude, and this trend suggests that the aryl group was migrating to an electron-deficient carbon. In this reaction $InCl_3$ actually activates the alkynes by coordinating its neighboring carbonyl group and thus lowers the LUMO [69] of the alkyne moiety (instead of α-diazocarbonyl compounds that traditionally occurred in diazo chemistry).

4.1.3 Triazole Derivatives

Recently, 1,2,3-triazolo-heterocyclic compounds, in particular sugar analogues having triazole moiety, have been attracting much attention in medicinal chemistry. This is due to the fact that oligosaccharides play important roles in many cellular functions and some of these classes of compounds have been reported to possess interesting inhibitory activities against glycosidases. Also the 1,2,3-triazole heterocyclic entity is an interesting moiety in terms of biological properties such as antibacterial, anti-HIV and antiallergic agents [70–74].

Various bicyclic 1,2,3-triazolo-heterocyclic compounds have been prepared by tandem azidation and the intramolecular 1,3-dipolar cycloaddition reaction of ω,ω-dialkoxy acetylenic compounds having an aliphatic acetal moiety with $TMSN_3$ in the presence of 1–5 mol % of $In(OTf)_3$. Accordingly reaction of dimethylacetal containing ynones **21a–f** with $TMSN_3$ **22** in the presence of 5 mol % $In(OTf)_3$ in 1,2-dichloromethane at 50 °C in 3–48 h afforded the corresponding triazolopyridine derivatives **23a–f** in moderate to good yields (60–93%). Other Lewis acids $Sc(OTf)_3$ and $Bi(OTf)_3$ were equally effective whereas $In(OTf)_3$ and $Yb(OTf)_3$ were less effective. In order to show the efficacy of this reaction, different substitution patterns including ethyl ester, benzoyl, trimethyl silyl and bromodifluoromethyl on the ynone moiety were studied and the results are summarized in Table 6 (Scheme 8) [75].

Table 6 $InCl_3$ catalyzed synthesis of triazoles

Ynones	R	Temperature (°C)	Time (h)	Product	Yield (%)
21a	COPh	50	3	**23a**	75
21b	$CO(CH_2)_2Ph$	50	4	**23b**	81
21c	CO_2Et	50	9	**23c**	72
21d	H	70	48	**23d**	72
21e	$SiMe_3$	70	20	**23e**	93
21f	CF_2Br	rt	36	**23f**	60

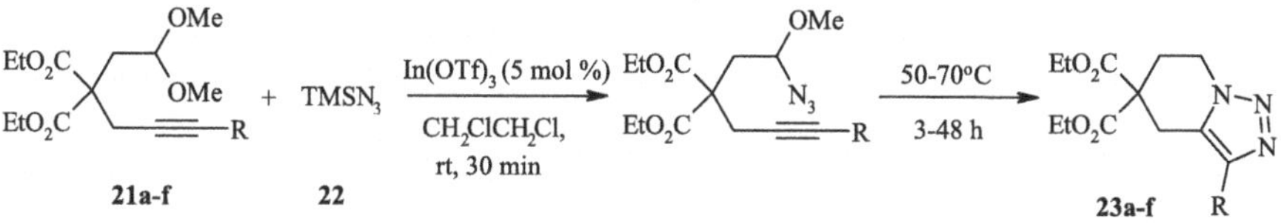

Scheme 8 Synthesis of triazolopyridines

Similarly, ynones with ether-tethered diethylacetals **24a–b** reacted with $TMSN_3$ **22** in the presence of 5 mol % $In(OTf)_3$ in 1,2-dichloromethane at 50 °C in 2–17 h to afford the corresponding triazolooxazine derivatives **25a–b** in good yields (75–83%) (Scheme 9).

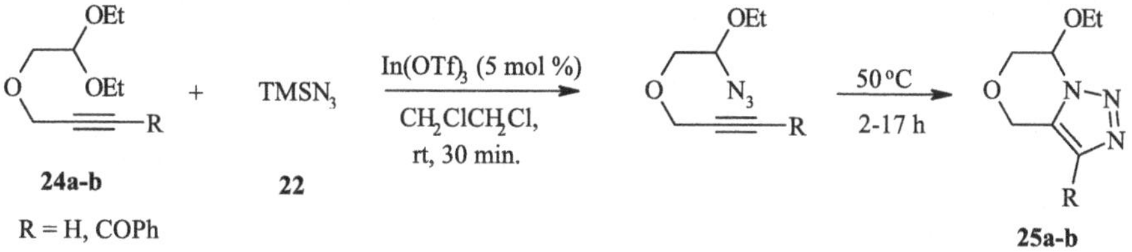

Scheme 9 Synthesis of triazolooxazines

Also, ynones with amino-tethered diethylacetals **26a–c** reacted with $TMSN_3$ **22** in the presence of 5 mol % $In(OTf)_3$ in 1,2-dichloromethane at 50 °C in 4–12 h to afford the corresponding triazolopyrazine derivatives **27a–c** in moderate to good yields (39–87%) (Scheme 10).

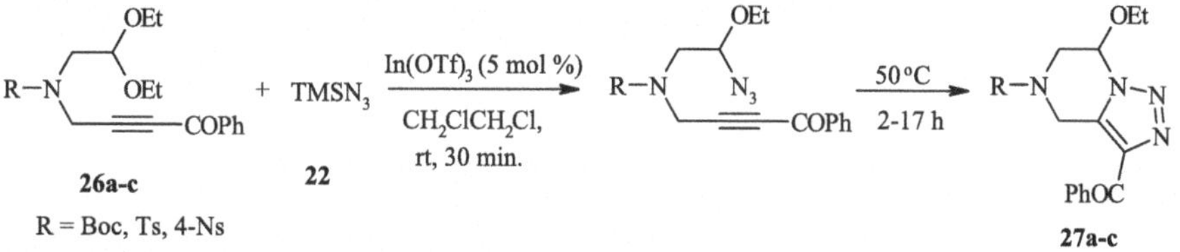

Scheme 10 Synthesis of triazolopyrazines

4.1.4 Pyrrolidines and Tetrahydrofurans

Nitrogen and oxygen-containing five-membered heterocycles such as pyrrolidines, including proline and related amino acids, and tetrahydrofurans are important structures because of their presence in many natural products and their biological activity [76, 77].

Methylenepyrrolidines **30a–e** have been prepared via [3+2] cycloadditions of ethenetricarboxylates **28a–e** with propargylamines **29a–b** in the presence of 1.2 eq of $InBr_3$ in dichloromethane at room temperature for 15–17 h. Similar results were also obtained in the presence of 0.2 eq of $InBr_3$-Et_3N in 1,2-dichlorethane at 80 °C for 4 h (Scheme 11) [78].

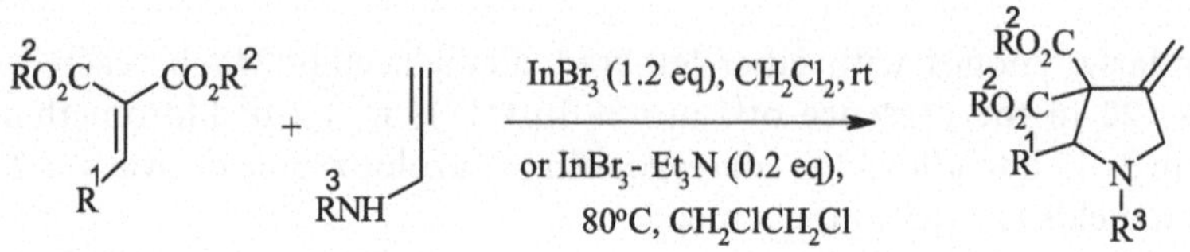

28a-e **29a-b** **30a-e**

Scheme 11 Synthesis of pyrrolidines

Other Lewis acids $ZnBr_2$, $ZnCl_2$, $Zn(OTf)_2$, $InCl_3$, $In(OTf)_3$ were equally effective for this reaction. This method requires no excess amount of the propargylic substrates. In all the cases the reactions proceeded very smoothly and the results are shown in Table 7.

Table 7 $InBr_3$ catalyzed synthesis of pyrrolidines

R^1	R^2	R^3	Reaction conditions	Catalysts	Product	Yield (%)
CO_2Et	Et	Me	CH_2Cl_2, rt	$InBr_3$	**30a**	64
			CH_2ClCH_2Cl, 80 °C	$InBr_3$–Et_3N		64
COPh	Et	Me	CH_2Cl_2, rt	$InBr_3$	**30b**	68
			CH_2ClCH_2Cl, 80 °C	$InBr_3$–Et_3N		69
CON–$(CH_2)_5$–	Et	H	CH_2Cl_2, rt	$InBr_3$	**30c**	69
H	^{t}Bu	Me	CH_2Cl_2, rt	$InBr_3$	**30d**	63
			CH_2ClCH_2Cl, 80 °C	$InBr_3$–Et_3N		73
H	^{t}Bu	H	CH_2Cl_2, rt	$InBr_3$	**30e**	100
			CH_2ClCH_2Cl, 80 °C	$InBr_3$–Et_3N		100
Me	Et	Me	CH_2Cl_2, rt	$InBr_3$	**30f**	73
			CH_2ClCH_2Cl, 80 °C	$InBr_3$–Et_3N		46

Similarly, methylenetetrahydrofurans **32a–c** have been prepared via [3+2] cycloadditions of ethenetricarbonyl derivatives **28a–c** with propargyl alcohols **31** in the presence of 1.2 eq of $InBr_3$ in dichloromethane at room temperature for 15–17 h. The similar results were also obtained in the presence of 0.2 eq of $InBr_3$-Et_3N in 1,2-dichlormethane at 80 °C for 4 h. This method fails with *tert*-butyl esters, probably because the *tert*-butyl cation generated in situ reacts with intermediates (Scheme 12) [79].

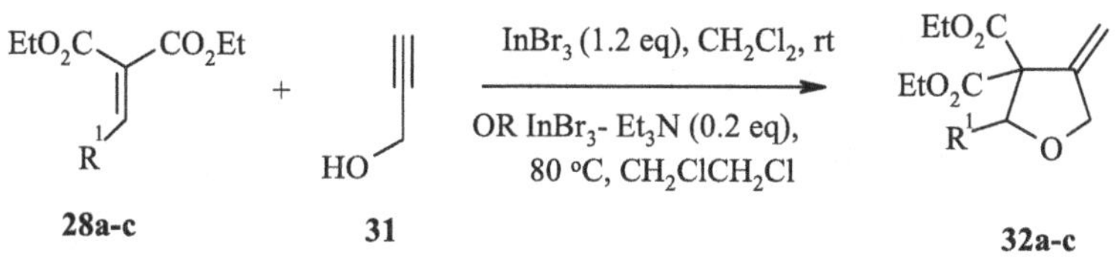

28a: R^1=CO_2Et
28b: R^1=COPh
28c: R^1=CON-$(CH_2)_5$-

Scheme 12 Synthesis of tetrahydrofurans

5 Six-Membered Rings

5.1 Carbo-Diels–Alder Reactions

The [4+2] Diels–Alder reaction is a conjugate addition reaction of a conjugated diene to an alkene or alkyne (the dienophile) to produce a cyclohexene moiety. The driving force of the reaction is the formation of new σ-bonds, which are energetically more stable than the π-bonds (Fig. 5). Since the reaction forms a cyclic product, via a cyclic transition state, it can also be described as a "cycloaddition." The reaction is a concerted process.

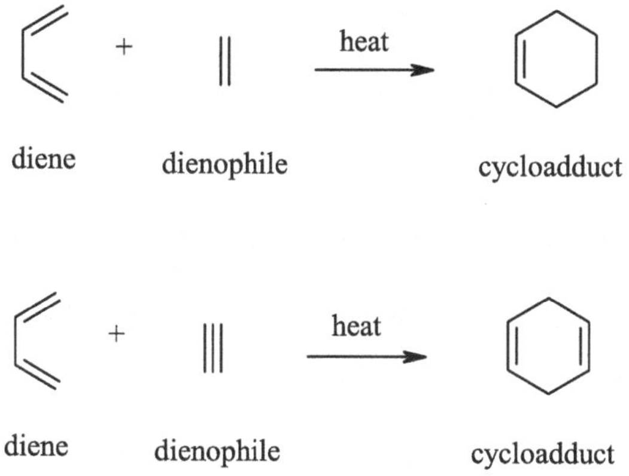

Fig. 5 Diels–Alder addition of ethylene and ethyne with butadiene

This [4+2] cycloaddition discovered by Diels and Alder in 1928 [80] is one of the most widely used methods of synthesis of six-membered rings [81]. The reaction follows either a synchronous and concerted mechanism or a multi-stage (non-concerted) and non-synchronous mechanism where the transition state can be a di-radical [82]. The mechanistic aspects of cycloadditions were well explained by the concept of conservation of orbital symmetry, developed by Woodward and Hoffmann [83]. Many different versions of the Diels–Alder reaction were described, including intramolecular [4+2] cycloadditions, hetero-Diels–Alder reactions, pressure-accelerated Diels–Alder reactions, and Lewis acid-accelerated Diels–Alder reactions.

A classical method to enhance regio- and stereoselectivity is based on the use of Lewis acid catalysts. Upon complexation of such species to the dienophile, the normal demand Diels–Alder reaction is promoted since the energy gap between the LUMO of the dienophile and the HOMO of the diene is reduced, thus decreasing the activation energy required to achieve the cycloaddition.

5.1.1 Epoxy Isoindolines

Epoxyisoindoline derivatives **36a–h** have been synthesized by the intramolecular Diels–Alder reaction of allyl(2-furfuryl)anilines **35a–h** (formed in situ by the reaction of (2-furfuryl)anilines **33a–h** with allyl bromide **34**) in the presence of 10 mol % of $In(OTf)_3$ under solvent-free conditions and microwave irradiation with excellent yields (80–90%) (Scheme 13).

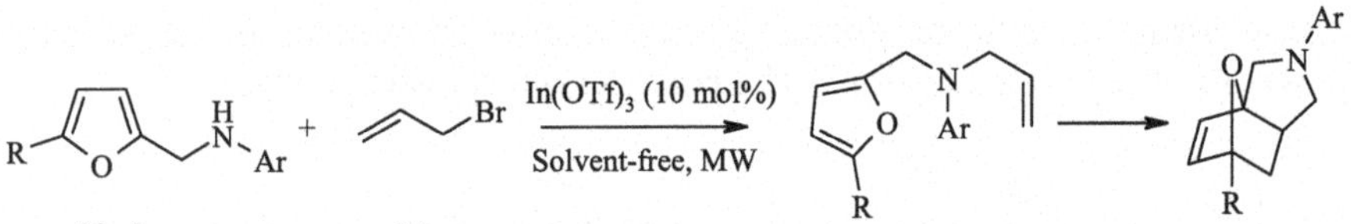

33a-h **34** **35a-h** **36a-h**

Scheme 13 Synthesis of epoxyisoindolines

Similarly tricyclic epoxyisoindoline **38** has been synthesized by the reaction of (2-furfuryl)-aniline **33a** with cyclohexenyl bromide **37** (Scheme 14). The advantage of this method was that the catalyst could be recovered and reused for subsequent runs without significant loss of the catalytic activity. The results are presented in Table 8 [84].

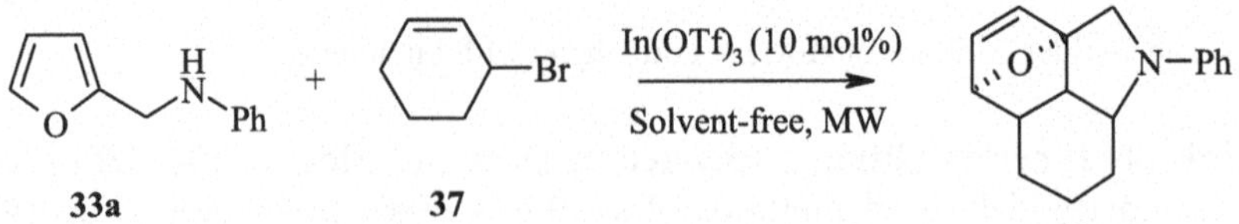

33a **37** **38**

Scheme 14 Synthesis of tricyclic epoxyisoindoline

Table 8 $In(OTf)_3$ catalyzed synthesis of epoxyisoindolines

Ar	R	Reaction time (min)	Product	Yield (%)
C_6H_5	H	8	**36a**	85
m-ClC_6H_4	H	8	**36b**	82
p-$C_2H_5OC_6H_4$	H	10	**36c**	85
p-$CH_3OC_6H_4$	H	9	**36d**	90
p-$CH_3C_6H_4$	H	8	**36e**	80
p-BrC_6H_4	H	8	**36f**	80
m-$CH_3OC_6H_4$	H	9	**36g**	82
p-$CH_3C_6H_4$	NO_2	10	**36h**	80

5.1.2
Isochromenones and Isoquinolinones

The intramolecular Diels–Alder reaction of 1,7,9-decatrienoate derivatives **39a–f** in the presence of 20 mol % $In(OTf)_3$ in H_2O-^iPrOH (6:1) at 70–80 °C for 8–24 h afforded the corresponding cycloadducts, hexahydrochromenone derivatives **40a–f**, in moderate yields. The results are presented in Table 9 (Scheme 15).

Table 9 $In(OTf)_3$ catalyzed synthesis of hexahydrochromenones

R^1	R^2	R^3	R^4	R^5	Temp (°C)	Time (h)	Product	Yield (%)
H	H	H	H	H	70	8	**40a**	82
Me	H	H	H	H	70	8	**40b**	76
H	H	Me	H	H	70	12	**40c**	83
H	H	H	Me	H	80	24	**40d**	68
H	H	H	^nPr	H	reflux	24	**40e**	18
H	H	H	H	COOH	rt	24	**40f**	78

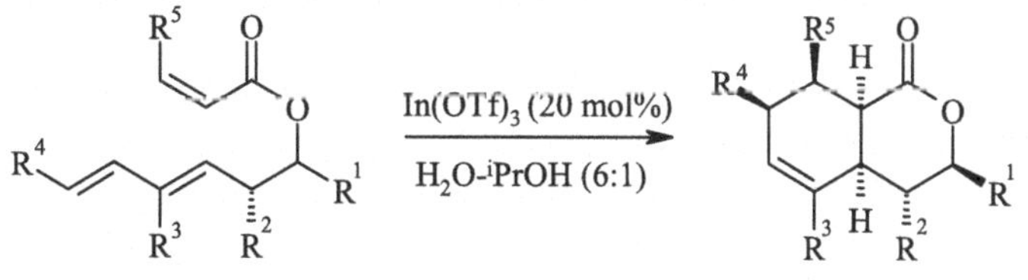

39a-f **40a-f**

Scheme 15 Synthesis of hexahydrochromenones

In the majority of cases the *endo* isomers formed as the major isomers. Similarly acrylamide derivative **41** underwent IMDA in the presence of 20 mol% $In(OTf)_3$ at 50 °C to afford the corresponding *endo*-hexahydro isoquinolinone **42** in 74% yield (Scheme 16) [85].

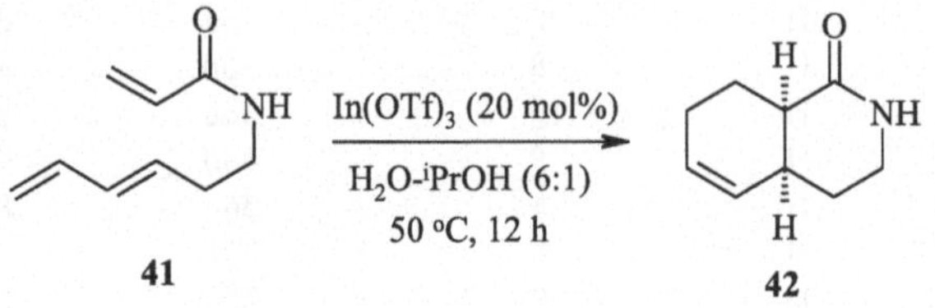

Scheme 16 Synthesis of hexahydro isoquinolinone

Other Lewis acids $InCl_3$, $Sc(OTf)_3$, $Yb(OTf)_3$, $Gd(OTf)_3$, $Ho(OTf)_3$, $Cu(OTf)_2$, $Zn(OTf)_2$ and AgOTf were less effective for this reaction. The advantageous aspects of this method were the use of eco-friendly solvents such as water and isopropanol, and the recovery and reusability of the catalyst.

5.2 Imino-Diels–Alder Reactions

Natural products with functionalized six-membered heterocyclic rings containing nitrogen are most abundant in nature and show significant biological activities. Most of these molecules show antibacterial activity, antitumor activity, antiparasitic and cytotoxic activity, as well as inhibition of bacterial cell division through a mechanism involving DNA alkylation [86, 87]. Among the various reactions to synthesize these heterocycles, the imino-Diels–Alder provides a rapid means of construction of functionalized heterocyclic rings with control of regio-, diastereo- and enantioselectivity [88–91].

The imino-Diels–Alder reaction adopts mainly three basic variants (Fig. 6). In the first and most common method the imine function appears as the dienophile and reacts with an electron-rich diene to form the corresponding cycloadduct. In the second and third variants the imine is found to be in the diene as either 1-azadiene or 2-azadiene structures, which reacts with an activated dienophile to form the corresponding adducts. The reaction between 2-azadiene and an activated dienophile is generally referred to as the Povarov reaction [92].

A limitation to the application of azadienes relative to the more commonly used oxodienes is that azadienes generally require some form of activation to achieve general synthetic utility. With appropriate substitution both 1-aza-1,3-butadienes and 2-aza-1,3-butadienes are reactive as either electron-rich or electron-deficient Diels–Alder dienes. As in the case of carbo-Diels–Alder reactions, imino-Diels–Alder reactions involve thermal activation as well as coordination of heteroatoms with Lewis acids, which provide excellent stereo-

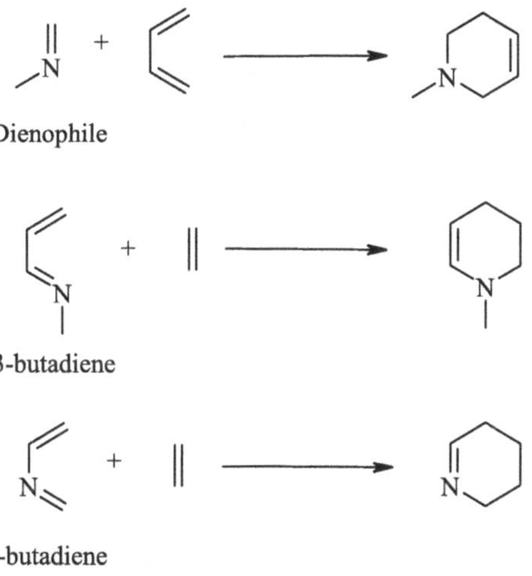

Fig. 6 Three basic variants in imino-Diels–Alder reactions

and regioselectivities. The theoretical and experimental evidence (vide supra) shows that cycloaddition with imino dienophiles can be concerted or stepwise. The stepwise mechanism involves a tandem Mannich–Michael reaction closely related to the aza-Diels–Alder reaction and has been included in this review for complete coverage. It is worth noting that reviews featuring the general synthetic utility of azadienes and their utility in the Diels–Alder reaction have appeared recently [93–97].

5.2.1 Cyclopentaquinolines

Cyclopentaquinolines **44a–g** have been prepared using a catalytic amount of anhydrous indium trichloride (20 mol %) by the reaction of various Schiff

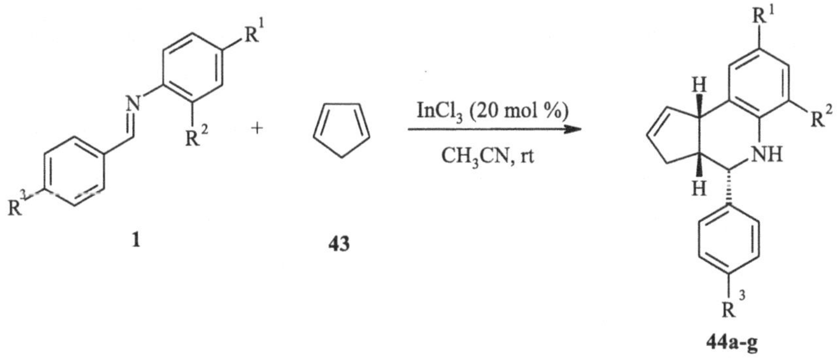

Scheme 17 Synthesis of cyclopentaquinolines

Table 10 $InCl_3$ catalyzed synthesis of cyclopentaquinolines

Schiff bases (1) R^1	R^2	R^3	Time (min)	Product	Yield (%)
H	H	H	30	**44a**	75
NO_2	H	H	30	**44b**	95
OCH_3	H	H	45	**44c**	58
Cl	H	H	30	**44d**	84
Cl	CH_3	H	30	**44e**	90
H	COOH	H	45	**44f**	78
H	C_2H_5	H	45	**44g**	82

bases **1** with cyclopentadiene **43** (Scheme 17) [98]. In this reaction Schiff bases **1** acted as azadienes and underwent [4+2] hetero-Diels–Alder reactions with cyclopentadiene **43** to afford the corresponding cyclopentaquinolines **44a–g** with high *endo* selectivity in 58–95% yield (Table 10). $InCl_3$ also catalyzes the hetero-Diels–Alder reaction of *N*-benzylidene-1-naphthylamine **1a** with cyclopentadiene **43** to afford benzo[*h*]-cyclopentaquinoline **45** in 75% yield (Scheme 18) [99].

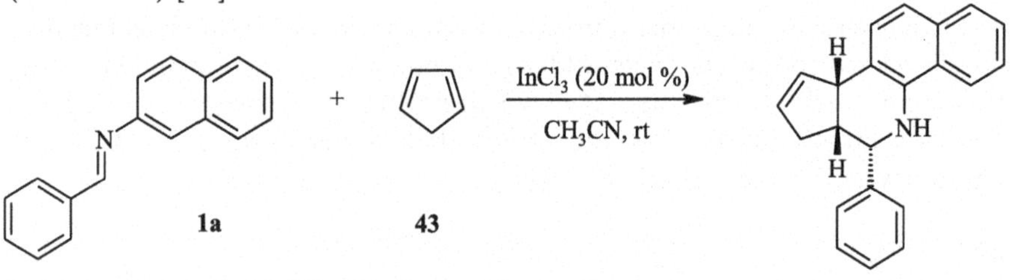

1a **43** **45**

Scheme 18 Synthesis of benzo[*h*]-cyclopentaquinoline

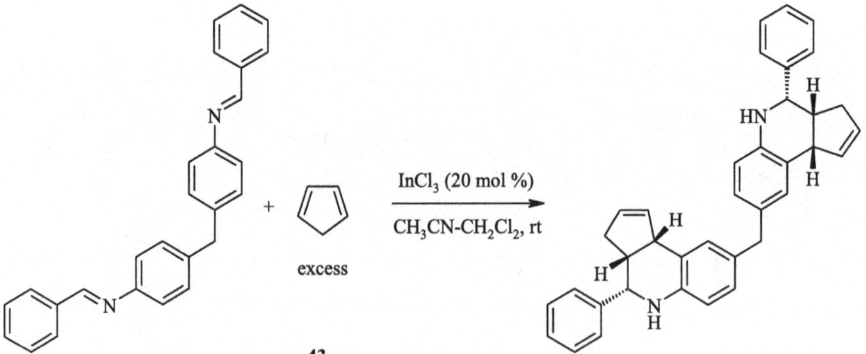

1b **43** **46**

Scheme 19 Synthesis of biscyclopentaquinoline

In a similar fashion, Schiff base **1b** derived from 4,4′-diaminodiphenylmethane reacted with excess cyclopentadiene **43** in the presence of $InCl_3$ to furnish the corresponding biscyclopentaquinoline **46** (Scheme 19) [100].

5.2.2 Indenoquinolines

Indenoquinoline derivatives were known to possess a wide range of biological activities such as 5-HT-receptor binding activity, antiinflammatory activity, antitumor activity and as an inhibitor for steroid reductase [101–104].

Table 11 $InCl_3$ catalyzed synthesis of indeno[2,1-*c*]quinolines

N-benzylidine anilines (1) R^1	R^2	R^3	Time (h)	Product	Yield (%)
H	H	H	6	**48a**	40
H	Cl	H	6	**48b**	48
H	OCH_3	H	24	**48c**	30
H	CH_3	H	12	**48d**	48
H	Cl	CH_3	9	**48e**	65

Imino-Diels–Alder reaction of *N*-benzylidine anilines **1** with indene **47** in the presence of 20 mol % indium trichloride afforded only the *endo* isomers of indeno[2,1-*c*]quinolines (**48a–e**) in moderate yields (40–65%) (Scheme 20 and Table 11) [105].

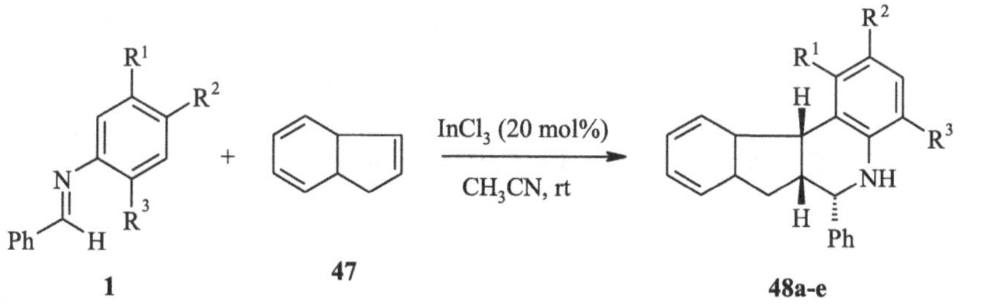

Scheme 20 Synthesis of indeno[2,1-*c*]quinolines

In the absence of indium trichloride, the reaction does not proceed at all. The *endo* isomer was obtained as a result of the likely secondary orbital interactions of diene and dienophile. Similarly, *N*-benzylidene-1-naphthylamine **1a** also reacted with indene **47** to provide benzo[*h*]-indeno[2,1-*c*]quinoline **49** in 50% yield (Scheme 21) [99, 106].

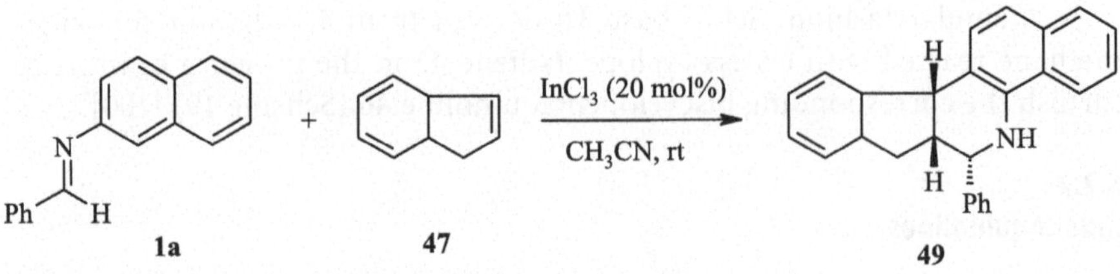

Scheme 21 Synthesis of benzo[*h*]-indeno[2,1-*c*]quinoline

5.2.3
Indolylquinolines

The indole system occurs in numerous natural products as well as in many therapeutic agents [107]. Compounds containing both the indole and the quinoline moiety possess a broad spectrum of biological activity [108, 109].

Table 12 $InCl_3$ catalyzed synthesis of indolylquinolines

Indolylimines	Substituents R^1	R^2	R^3	Time (h)	Product	Yield (%)
1c	H	H	H	0.5	**50a**	83
1d	H	Me	H	0.5	**50b**	60
1e	H	Cl	H	0.5	**50c**	85
1f	Me	H	Me	0.5	**50d**	87
1g	Me	Cl	H	0.5	**50e**	79

Imino-Diels–Alder reaction of indolylimines **1c–g** with cyclopentadiene **43** in the presence of 20 mol % $InCl_3$ or 5 mol % indium triflate furnished the corresponding indolylquinoline derivatives **50a–e** in good yields (60–87%) [99, 110] and the results are presented in Table 12 (Scheme 22).

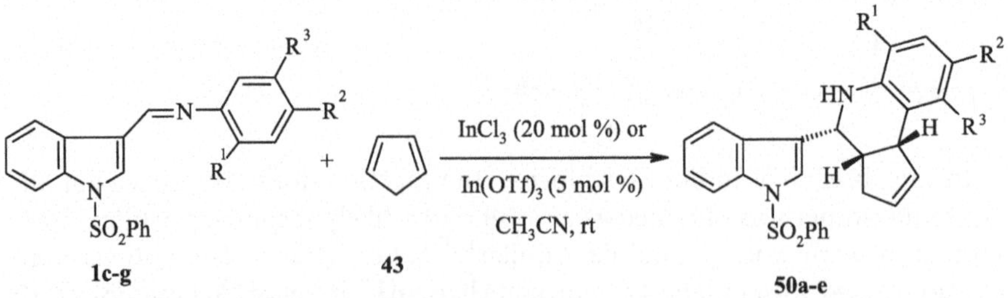

Scheme 22 Synthesis of indolylquinolines

The structures of the products were ascertained by the single crystal X-ray analysis of **50a** and **50c** [111].

5.2.4
Quinolino Pyrimidinones

Five-substituted pyrimidinone derivatives are known for their antitumor, antiviral, antitubercular, antifungal, molluscidal and larvacidal activity. Such compounds have shown activity against positive strand (vesicular stomatitis virus) RNA virus [112–117]. Quinolino pyrimidinones **53a–l** have been synthesized via imino-Diels–Alder cycloaddition reactions of *N*-arylimines **1** (2-azadiene) with unactivated 1,3-butadiene-tethered pyrimidinones **51a–c** in the presence of 15 mol% of $InCl_3$ in CH_2Cl_2 at room temperature in moderate yields (11–78%). These quinoline-substituted pyrimidinones

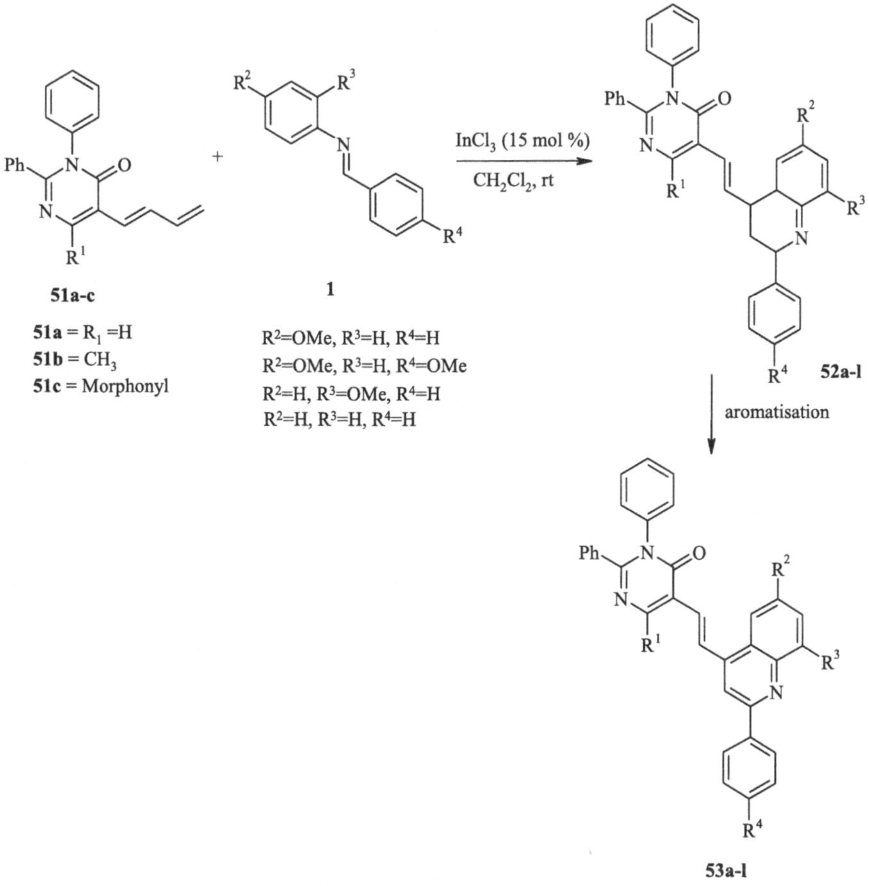

Scheme 23 Synthesis of quinilino pyrimidinones

have been formed via oxidation of the initially formed intermediate **52** (Scheme 23) [118].

In this aza-Diels–Alder reaction 2-azadiene acts as 4π component, while the dienyl pyrimidinone acts as 2π component. Other Lewis acids $MgBr_2$, $ZnCl_2$, $Yb(OTf)_3$ and $Sc(OTf)_3$ were also efficient catalysts for this transformation. However the reaction between **51a–c** and **1** ($R^2 = R^3 = R^4 = H$) did not proceed with the above mentioned catalysts. When mixed Lewis acid system $MgBr_2$/cat. $AlCl_3$ was used for this particular conversion, quinolino pyrimidinones **53i–l** were obtained in 74–86% yield.

5.2.5 Benzothiophenyl Cyclopentaquinolines

The imino-Diels–Alder reaction of benzo[*b*]-thienylimines **1h–k** with cyclopentadiene **43** in the presence of 20 mol% anhydrous $InCl_3$ provided benzo[*b*]thiophenyl cyclopentaquinoline derivatives **54a–d** in good yields with excellent *endo* stereoselectivities (81–82%) (Scheme 24) and the results are shown in Table 13 [119].

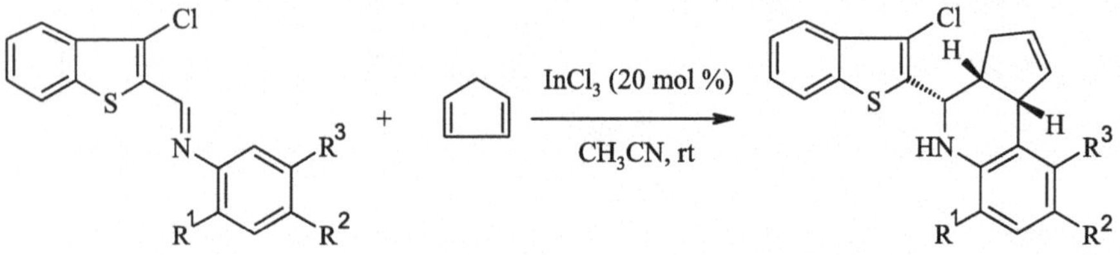

1h-k **43** **54a-d**

Scheme 24 Synthesis of benzo[*b*]thiophenyl cyclopentaquinolines

Table 13 $InCl_3$ catalyzed synthesis of benzo[*b*]thiophenyl cyclopentaquinolines

Schiff base	Substituents R^1	R^2	R^3	Product	Yield (%)
1h	CH_3	Cl	H	**54a**	85
1i	H	CH_3	H	**54b**	82
1j	H	Cl	H	**54c**	92
1k	H	H	H	**54d**	81

5.2.6 Benzothiophenyl Pyranoquinolines

The imino-Diels–Alder reaction of benzo[*b*]-thienylimines **1i,k** with 3,4-dihydro-2*H*-pyran **55** in the presence of 20 mol% anhydrous $InCl_3$ provided

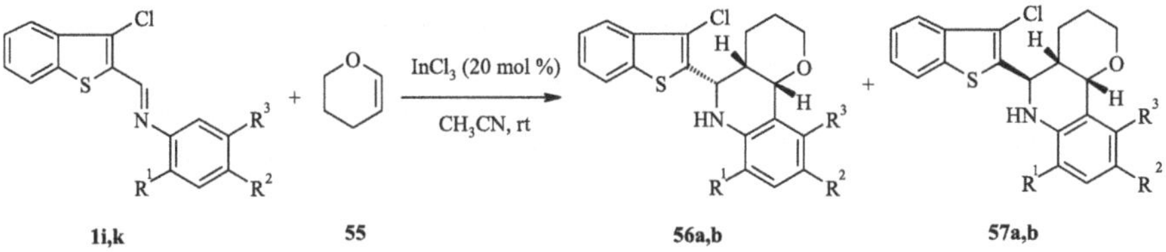

Scheme 25 Synthesis of benzo[*b*]thiophenyl pyranoquinolines

diastereomeric benzo[*b*]thiophenyl pyranoquinoline derivatives in moderate yields [119]. The diastereomers **56a** and **57a** were obtained in a ratio 68:32 in an overall yield of 57%, where **56b** and **56b** were obtained in a ratio of 46:54 in an overall yield of 55% (Scheme 25).

5.2.7 Pyranoquinolines

Pyranoquinoline derivatives are an important class of natural products and exhibit a wide spectrum of biological activities, such as antiallergic, antiinflammatory, antipyretic, analgesic, antiplatelet, psychotropic and estrogenic activity [120–123]. Many biologically active alkaloids, such as simulenoline, huajiaosimuline, zanthodioline, flindersine are known to possess the pyranoquinoline moiety (Fig. 7) [122, 123].

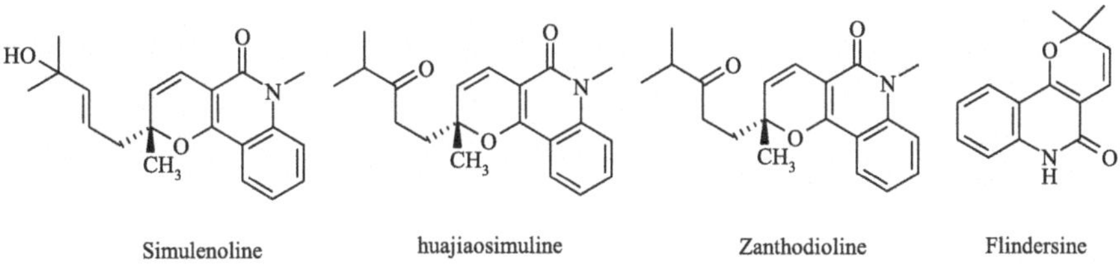

Fig. 7 Biologically important pyranoquinolines

Imino-Diels–Alder reaction of *N*-benzylidine anilines **1** with 3,4-dihydro-2*H*-pyran **55** in the presence of 20 mol % $InCl_3$ resulted in the corresponding pyrano[3,2-*c*]quinolines **58a–f** (*cis*) and **59a–f** (*trans*) in 45–80% overall yields and the results are summarized in Table 14 (Scheme 26) [106]. $InCl_3$ was the best choice among the other catalysts acetic acid, $BF_3{\cdot}OEt_2$ in terms of yields and selectivites.

$InCl_3$ also effectively catalyzes the imino-Diels–Alder reaction of 3,4-dihydro-2*H*-pyran **55** with *N*-benzylidine 1-naphthylamine **1a** to furnish the corresponding benzo[*h*]pyranoquinoline derivatives **60** (*cis*) and **61** (*trans*) in a ratio of 32:68 in an overall yield of 53% (Scheme 27) [106].

Table 14 $InCl_3$ catalyzed synthesis of pyrano[3,2-*c*]quinolines

Entry	Schiff bases (1) R^1	R^2	R^3	Time (min)	Product ratio *cis*:*trans* (58:59)	Overall yield (%) (58+59)
a	H	H	H	30	41:59	80
b	H	Cl	H	30	34:66	50
c	H	OCH_3	H	240	58:42	70
d	H	CH_3	H	120	68:32	70
e	H	Cl	CH_3	40	5:95	46
f	CH_3	H	CH_3	180	5:95	45

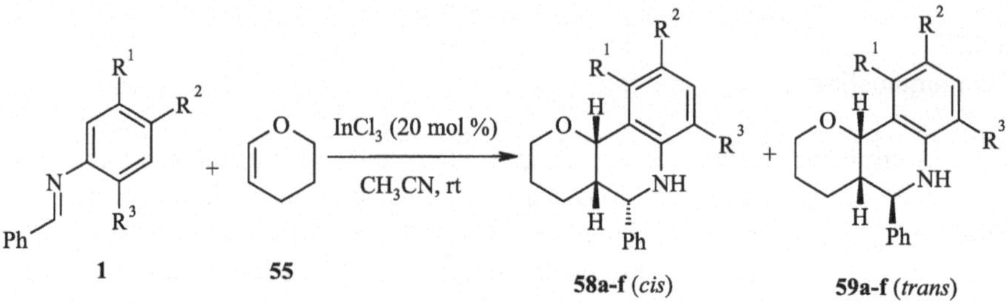

Scheme 26 Synthesis of pyrano[3,2-*c*]quinolines

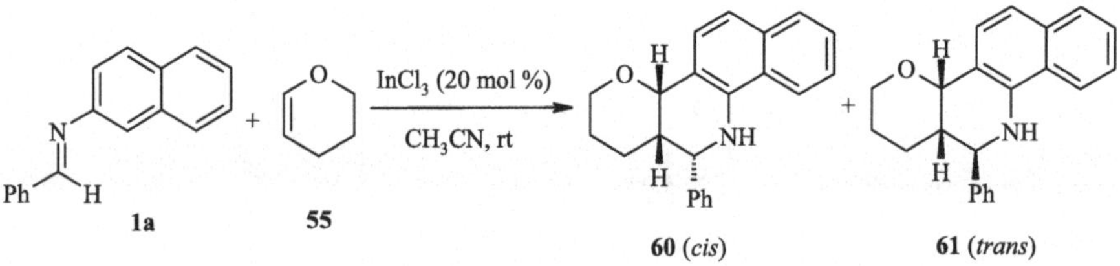

Scheme 27 Synthesis of benzo[*h*]pyranoquinolines

Similarly, imino-Diels–Alder reaction of imines **63a–k** (generated in situ from aryl amines **62** and 3,4-dihydro-2*H*-pyran **55**) with 3,4-dihydro-2*H*-pyran **55** in the presence of 5 mol% $InCl_3$ afforded the corresponding pyrano[3,2-*c*]quinolines **64a–k** (*endo*) and **65a–k** (*exo*) in 70–90% overall yields. All the reactions proceeded smoothly in high yields at ambient temperature. In most of the cases, the products were obtained as a mixture of *endo* and *exo* isomers, favoring the *endo* diastereomers and the results are summarized in Table 15 (Scheme 28) [124].

The product ratio was determined by ^1H NMR spectrum of the crude product. The stereochemistry of the product **64a** was assigned on the basis of coupling constants and nOe studies. The two six-membered quinoline

Table 15 $InCl_3$ catalyzed synthesis of pyrano[3,2-*c*]quinolines

Aryl imine	R	Reaction time (h)	Yield (%)	Product ratio *endo*:*exo* (64:65)
63a	H	4.0	90	95:5
63b	4-Me	3.5	85	90:10
63c	2-Me	4.0	89	92:8
63d	4-OMe	3.5	87	85:15
63e	4-F	4.5	85	95:5
63f	4-Cl	5.0	90	93:7
63g	4-Br	4.5	88	90:10
63h	2,3-F,F	6.0	80	85:15
63i	3-NO_2	7.0	78	70:30
53j	4-CH_2CN	4.5	89	98:2
63k	2-CN	6.0	70	60:40

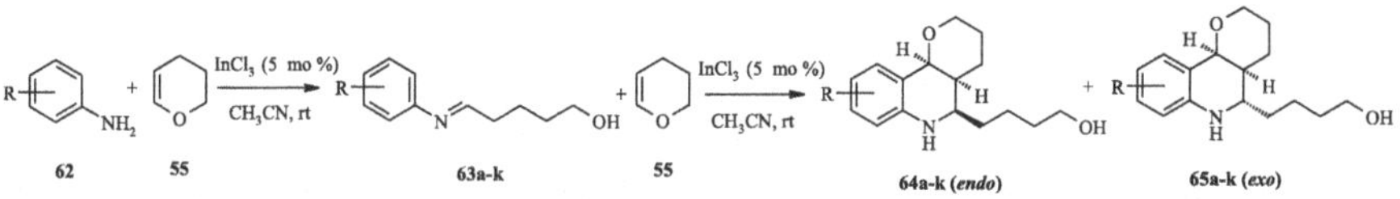

Scheme 28 Synthesis of pyrano[3,2-*c*]quinolines

and tetrahydropyran rings were *cis*-fused as depicted by the small coupling constant value $J_{H5-H6} = 5.6$ Hz for H_5 (δ5.0 ppm) proton as well as the observation of nOe cross peaks between them in the NOESY spectrum. The middle six-membered quinoline ring conformation was confirmed as a twist conformation, which was consistent with the small coupling constant value $J_{H6-H7} = 6.9$ Hz, for H_7 (δ3.34 ppm) and the presence of nOe cross peaks between H_5–H_6, H_6–H_7 in the NOESY spectrum. While the six-membered tetrahydropyran ring takes a chair conformation of $^5C_{13}$ (here 5 and 13 are the carbon atoms in the molecule), which was consistent with the large coupling constants values $J_{H14ax-H13ax} = 11.9$ Hz for H_{14ax} (δ3.39 ppm), $J_{H6-H12ax} = 12.1$ Hz for H_6 (δ2.00 ppm) as well as the presence of nOe cross peak between H_4–H_{14ax} (Fig. 8) and absence of nOe cross peak between H_5–

Fig. 8 Chemical structure and important nOes of compound **64a**

Fig. 9 Chemical structure and important nOes of compound **65a**

H_{14} in the NOESY spectrum (Fig. 9). In *exo* isomer **65a** as well as in *endo* isomer **64a**, the two six membered quinoline and tetrahydropyran rings were *cis*-fused as depicted by the small coupling constant value $J_{H5\text{-}H6} = 3.2$ Hz for H_5 (δ4.42 ppm) proton as well as the observation of nOe cross peak between them and the absence of nOe cross peak between H_6–H_7 in the NOESY spectrum [124].

5.2.8 Furanoquinolines

Furanoquinoline derivatives are an important class of natural products and exhibit a wide spectrum of biological activities, such as antimicrobial, antiviral, antiplasmodial, anti-HIV, anti-TB, mutagenic and cytotoxic activity. Many biologically active alkaloids, such as dictamine, flindersiamine, roxiamine A, and evolitrine are known to possess furanoquinoline moiety (Fig. 10) [125–128].

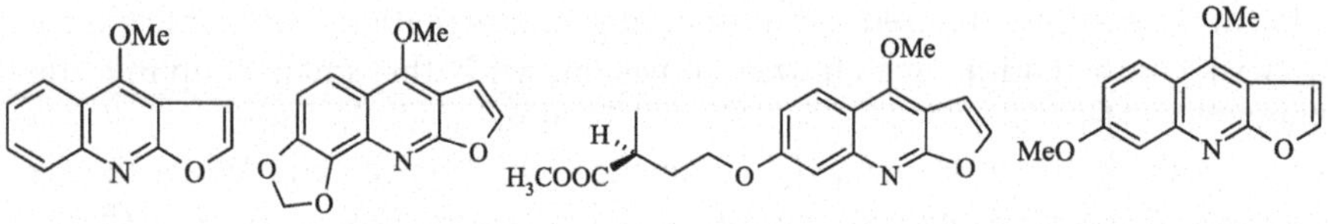

Dictamine Flindersiamine Roxiamine A Evolitrine

Fig. 10 Biologically important furanoquinolines

The imino-Diels–Alder reaction of imines **67a–h** with cyclic enol ether like 3,4-dihydro-2*H*-pyran **66** in the presence of 10 mol % $InCl_3$ in water as solvent afforded the diastereomeric furanoquinolines **68a–h** (*endo*) and **69a–h** (*exo*) in 46–84% overall yield. The best results were obtained by increasing the reaction temperature to 50–60 °C. In most of the cases cyclization products show *endo* selectivity. Imines bearing electron-donating groups were found to be more reactive than the ones bearing electron-withdrawing groups and the results are summarized in Table 16 (Scheme 29) [129].

The diastereoselectivity of the cyclization could be rationalized by a balance of steric effect (**A**) and chelation effect (**B**). Whereas a steric effect favors

Table 16 $InCl_3$ catalyzed synthesis of furanoquinolines

Aryl imine	R	Reaction conditions	Product ratio *endo:exo* (68:69)	Overall yield (%) (68+69)
67a	H	rt/48 h	78:22	65
67b	Me	rt/4 h	81:19	84
67c	OMe	rt/4 h	87:13	81
67d	Br	45 °C/10 h	87:13	81
67e	F	rt/10 h	86:14	81
67f	OH	rt/2 h	96:4	73
67g	NHPh	rt/10 h	86:14	65
67h	CN	rt/24 h	69:31	46

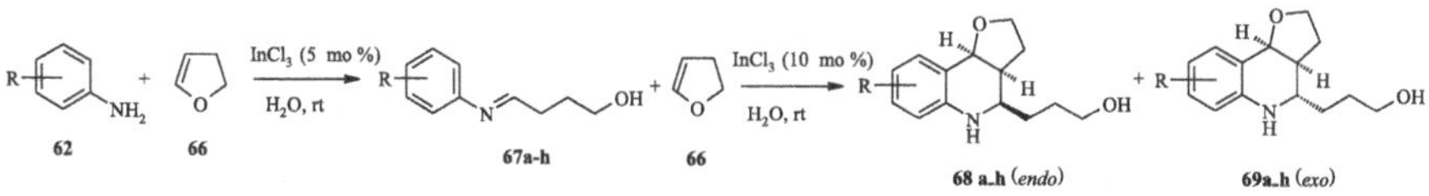

Scheme 29 Synthesis of furanoquinolines

the *exo* diastereomer, the chelation of indium(III) ion with the imine and the enol ether favors the *endo* isomer (Fig. 11).

Fig. 11 Proposed transition states *A* and *B* for the formation quinolines

This hypothesis is consistent with the observed switch of selectivity for entries 6 and 10 (Table 16). The presence of Br– and CN– reduce the electron density of aniline (more effectively than the other substituents), which decreases the chelation ability of the corresponding imines. As the *p*-phenylamino group is most likely also coordinated with indium(III) (and thus becomes an electron-withdrawing group), the observed *trans* selectivity with entry 9 (Table 16) can be explained similarly. The use of five-membered cyclic enol ethers decreased the steric effect and thus increased the *endo* selectivity (Table 16).

A tentative mechanism to rationalize the product formation is shown in Fig. 12. The cyclic enol ether can be hydrated easily in the presence of protonic or Lewis acid in water to give **I** [130, 131]. Then **I** undergoes facile ring opening in the presence of the indium(III) ion to give **II** in water. The condensation reaction between anilines and **II** will generate imine **III**, which is coordinated with the indium ion. Finally, an imino-Diels–Alder reaction of the imine **III** with the cyclic enol ether will generate the quinoline derivatives (Fig. 12).

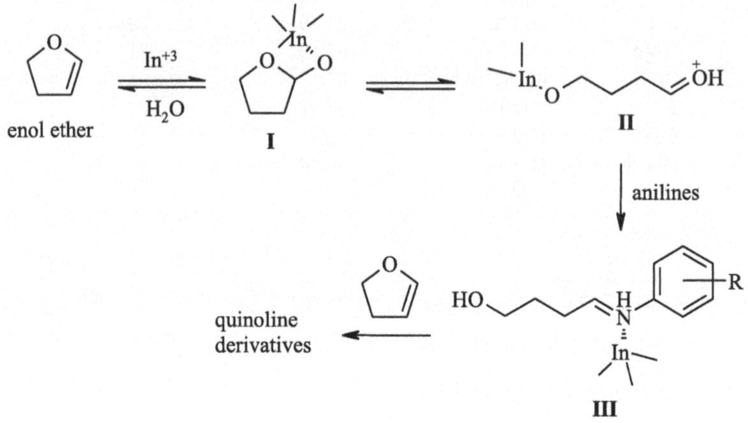

Fig. 12 Proposed mechanism for the formation quinolines

Interestingly, the reaction of anilines **62** and cyclic hemiacetal 2-hydroxy-tetrahydro-2*H*-pyran **70** in the presence of 10 mol % $InCl_3$ in water as solvent yielded the corresponding pyranoquinolines as a pair of diastereomers **64** (*endo*) and **65** (*exo*) in 28–85% overall yield. Also imino-Diels–Alder reaction of 2-chromanol **71** with aniline **62a** produced smoothly [2-(*o*-hydroxy-phenyl)]ethyl-substituted pyranoquinolines **72** (*endo*) and **73** (*exo*) (Scheme 30).

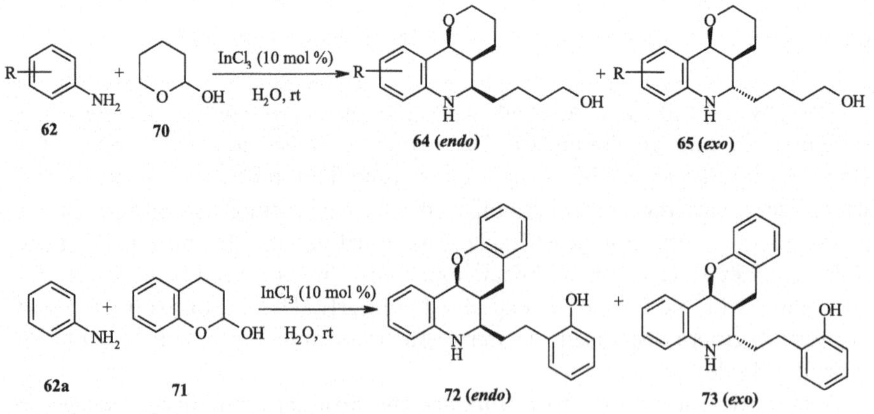

Scheme 30 Synthesis of pyranoquinolines

Polyaniline supported indium(III) trichloride (PANI-In) (10 mol %) also has been used as an efficient catalyst in imino-Diels–Alder reaction of cyclic enol ethers like 3,4-dihydro-2*H*-pyran **55** and 2,3-dihydrofuran **66** with aromatic amines **62** in water to afford the corresponding pyrano (**64** and **65**) and furanoquinolines (**68** and **69**) in good to excellent yields. The simple procedure of catalyst preparation, easy recovery and reusability of the catalyst makes this method an environmentally benign chemical process. The doping of polyaniline with $InCl_3$ is depicted in Scheme 31 [132].

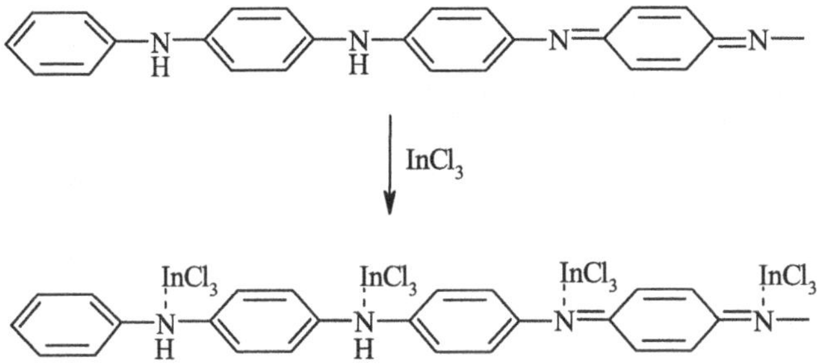

Scheme 31 Doping of polyaniline with $InCl_3$

5.2.9 Pyrroloquinolines

Pyrroloquinolines forms the core structural unit for a number of biologically interesting molecules. Antineoplastic agent, gastric (H^+/K^+) ATPase inhibitor and natural product, non-peptide Bradykinin inhibitor, matrinellic acids all possess pyrroloquinoline moiety in various oxidation states (Fig. 13) [133–136].

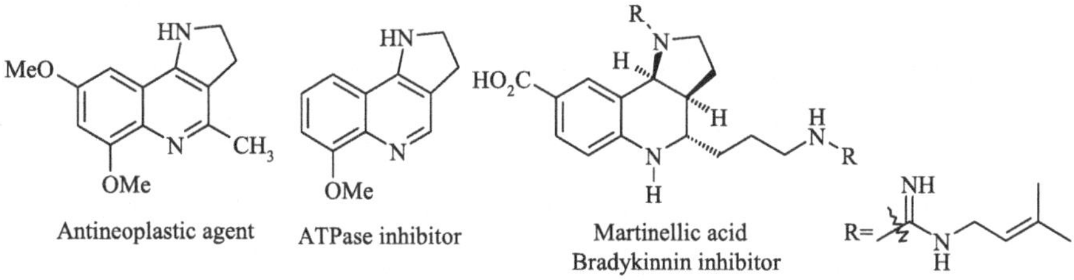

Fig. 13 Biologically important pyrroloquinolines

The imino-Diels–Alder reaction of imines **1** with cyclic enamide **74a** (readily prepared from the reaction of corresponding imine with propionyl chloride) took place rapidly at room temperature in the presence of 15 mol % anhydrous $InCl_3$ and provided diastereomeric pyrrolopyra-

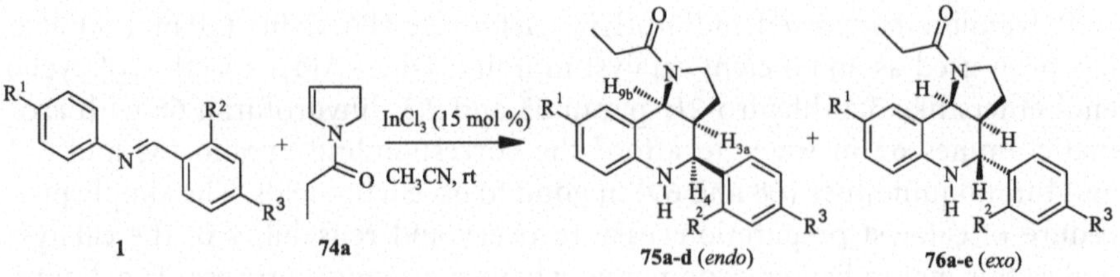

Scheme 32 Synthesis of pyrroloquinolines

noquinoline derivatives **75a–d** (*endo*) and **76a–d** (*exo*) in overall moderate yields (41–50%) (Scheme 32). In all the cases thermodynamically more stable *endo* isomers were the major isomers. When bulkiness of the phenyl group was increased by introducing an *o*-nitro substituent, then the amount of *endo* adduct **75c** substantially increased. This cycloaddition reaction works equally well for imines containing both electron-donating and electron-withdrawing groups *para* to the imine nitrogen. Scandium triflate and trifluoromethane sulphonic acid also catalyzed this reaction well. However, some hydrolysis of the imines makes these less ideal than $InCl_3$. However, the reaction failed when a strong electron-releasing group was present on the aldehyde end of the imine. The results are depicted in Table 17; it was found that the reaction was unsuccessful with imines derived from aliphatic aldehydes and methyl glyoxalate (< 10% yields). The reactions of imines **77a–b** derived from cinnamaldehyde and 3-cyanoacrolein provided moderate yields of corresponding pyrroloquinolines **78a–b** (*endo*) and **79a–b** (*exo*) favoring *endo* selectivity (Scheme 33) [137, 138].

Table 17 $InCl_3$ catalyzed synthesis of pyrroloquinolines

Aryl imines	R^1	R^2	R^3	Product ratio *endo*:*exo* (75:76)	Overall yield (%) (75+76)
1a	H	H	H	1:1	41
1b	OMe	H	H	1:0.8	50
1c	H	NO_2	H	2:1	33
1d	CO_2Me	H	H	1:1	48
1e	H	H	OMe	0	0

The stereochemistry at the ring junction in isomers **75a** and **76a** was readily assigned as *cis* due to the axial-equatorial coupling constants $J_{H3a\text{-}H9b}$ of 6.7 and 6.2 Hz, respectively. This was further confirmed by nOe difference spectroscopy, where saturation of proton H_{9b} gave enhancements of 6.0 and 6.8% onto H_{3a} for the *endo* and *exo* isomers, respectively. Assigning *exo*, *endo*

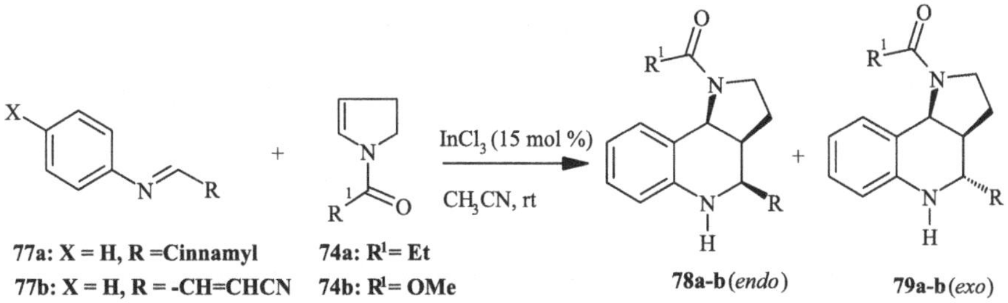

Scheme 33 Synthesis of pyrroloquinolines

stereochemistry to each of the isomers **75a** and **76a** proved more difficult than initially anticipated. This was because the *exo* isomers **76** adopted an unexpected conformation where the aryl group and the alkyl group of the pyrrolidine ring were *trans*-diaxial. Equatorial proton H_{3a}, was antiperiplanar to the electronegative secondary aromatic amine and this resulted in a further decrease in the axial-equatorial vicinal coupling constant J_{H3a-H4} [139]. As a result of these two effects, the vicinal coupling constant J_{H3a-H4} for the *endo* and *exo* isomer **75a** and **76a** were remarkably similar at 2.7 and 2.5 Hz, respectively. Clearly these isomers cannot be distinguished from measurement of vicinal coupling constant J_{H3a-H4}. Obtaining nOe difference experiments confirmed the stereochemistry of *endo* adduct **75a**. Hence, saturation of proton H_{9b} gave an enhancement to H_4 of 2% indicating these protons where *cis* and also 1,3-diaxial. With the corresponding *exo* isomer **76a** saturation of proton H_{9b} resulted in an nOe enhancement onto the protons of the pendent aromatic ring of 2.0%. This confirmed the stereochemistry and indicated that both proton H_{9b} and the phenyl group were 1,3-diaxial. This analysis of the stereochemistry and conformation in solution was further confirmed in the solid state by single crystal X-ray analysis of *endo* and *exo* adducts **75a** and **76a**, respectively [138].

However, the reaction of imine **1** with acyclic enamide **80** proceeded smoothly at room temperature to afford the corresponding pyrroloquinolines **81** (*endo*) and **82** (*exo*) in a 1:2 ratio in overall 67% yield. The stereochemistry of the products was confirmed by nOe experiments. The relative stereochemistry of the major *exo* isomer is 2*S*′3*R*′4*S*′ with the preserved stereochemistry

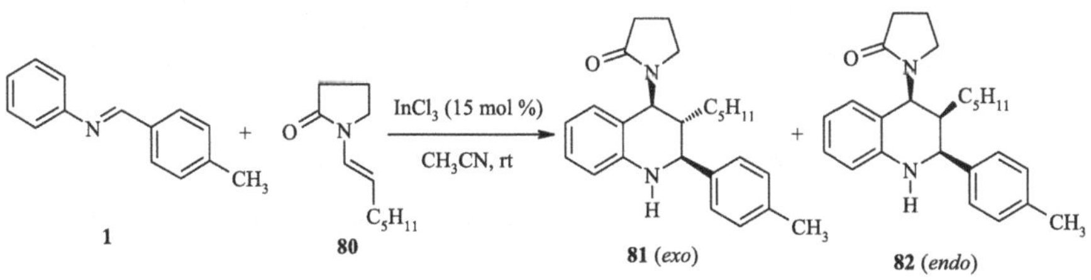

Scheme 34 Synthesis of pyrroloquinolines

of the enamide. Also, the stereochemistry of the minor *endo* isomer is 2S′3S′4S′ with the reversed relative stereochemistry of the enamide. These results clearly indicate that adduct **81** was formed by a non-concerted process (Scheme 34).

5.2.10 Chromanoquinolines

Diastereomeric tetrahydrochromanoquinolines **84a–i** (*cis*) and **85a–i** (*trans*) have been synthesized by the reaction of aromatic amines **62** and *O*-allyl derivatives of salicylaldehydes **83a–i** in the presence of 20 mol % $InCl_3$ via the intramolecular [4+2] imino-Diels–Alder cycloaddition (IMIDA) of imines (derived in situ by the reaction of anilines with aldehyde functionality of salicylaldehydes) with the olefinic moiety of the *O*-allyl group, in acetonitrile at room temperature in short reaction times with excellent overall yields (Scheme 35) (87–98%).

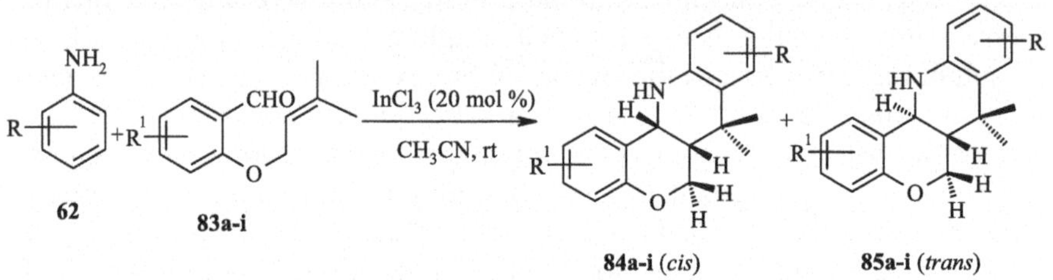

Scheme 35 Synthesis of tetrahydrochromanoquinolines

In all cases, the products were obtained as a mixture of *cis* and *trans* isomers in a 1:1 ratio, determined from the ^1H NMR spectrum of the crude product. These isomers were isolated by column chromatography on silica

Table 18 $InCl_3$ catalyzed synthesis of tetrahydrochromanoquinolines

Entry	Aromatic amines (62) R	R^1	Time (min)	Products *cis* (84) (%)	*trans* (85) (%)	Overall yield (%) (84+85)
a	H	H	5	52	46	98
b	H	3-CH_3O	5	45	49	94
c	2-CH_3	H	8	44	48	92
d	4-CH_3O	H	10	47	49	96
e	4-Br	H	10	45	48	93
f	4-NO_2	H	15	42	47	89
g	4-COOH	H	15	42	49	91
h	4-CN	H	10	46	41	87
i	H	Cl	5	50	44	94

gel. The *cis* and *trans* stereochemistry of the products was assigned on the basis of coupling constants of the protons in the ^{1}H NMR spectroscopy and the results were summarized in Table 18 [140].

5.2.11 Benzoanthracenes

The imino-Diels–Alder reaction of chromone Schiff bases **87a–f** (generated in situ by the reaction of aromatic amines **62** with 3-formyl-chromone **86**) with 3,4-dihydro-2*H*-pyran **55** in the presence of 10 mol % anhydrous $In(OTf)_3$ resulted in corresponding benzo[*a*]anthracenes **88a–f** in moderate yields (70–75%) and with exclusively *endo* stereoselectivity. Scandium triflate also worked well for this reaction. All reactions were carried out in one pot at room temperature. Furthermore, the reaction did not proceed in the absence of the catalyst (Scheme 36) [141].

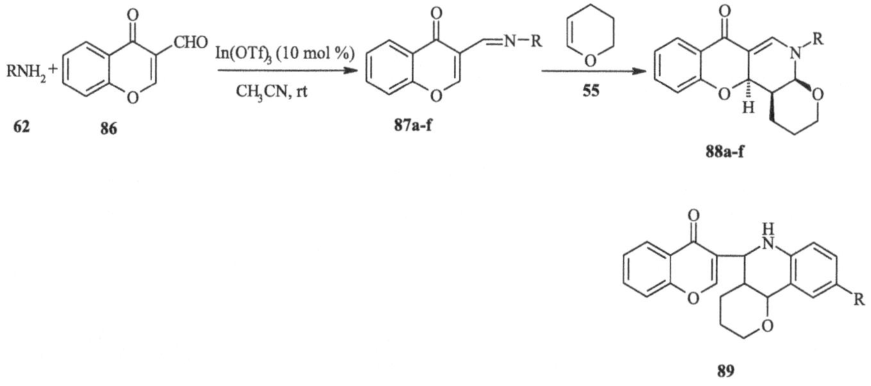

Scheme 36 Synthesis of benzo[*a*]anthracenes

Table 19 $InCl_3$ catalyzed synthesis of chromanoquinolines

Schiff base	R	Reaction time (h)	Product	Yield (%)
87a	C_6H_5	8	**88a**	80
87b	4-MeC_6H_4	8	**88b**	78
87c	4-$MeOC_6H_4$	9	**88c**	75
87d	4-BrC_6H_4	8	**88d**	90
87e	4-ClC_6H_4	8	**88e**	85
87f	$C_6H_5CH_2$	10	**88f**	55

The structural assignment for the cycloadduct **88a** was assigned as *endo* based on the basis of its high-resolution ^{1}H and ^{13}C spectra. The diagnostic signal for the proton H_a in **87a**, which appeared at δ8.65, was absent in the cycloadduct **88a**, whilst the upfield shift of the proton from δ8.65 to 5.34 (12a–H) showed that the cycloaddition had occurred at the C-2 position of the chromone unit, which clearly ruled out the possibility of any formation of product of type **89**, and also there was no evidence for the formation of any *exo* product. The observed values for the allylic coupling constants in compounds **88** are consistently $\approx$1 Hz and therefore invite assignment of these products as having *endo* stereochemistry. The results were summarized in Table 19.

5.2.12 Pyrazolo-Thiopyranoquinolines

Thiopyranoquinolines were reported as interleukin-1 inhibitors, additionally thiopyrazoles were known as a series of cox-2-selective inhibitors, which demonstrates an antiinflammatory activity [142–144].

Diastereomeric tetrahydropyrazolo[4′,3′:5,6]thiopyrano[4,3-*b*]quinolines **90a–j** (*cis*) and **91a–j** (*trans*) have been prepared by the reaction of aromatic amines **62** and *S*-allyl derivatives of pyrazoloaldehydes **89a–b** in the presence of 20 mol % $InCl_3$ via the intramolecular [4+2] imino-Diels–Alder cycloaddition (IMIDA) of imines (derived in situ by the reaction of anilines with the aldehyde functionality of pyrazoloaldehydes), with an olefinic moiety of the *S*-allyl group in acetonitrile at room temperature in short reaction times with excellent overall yields (85–95%) (Scheme 37) [145].

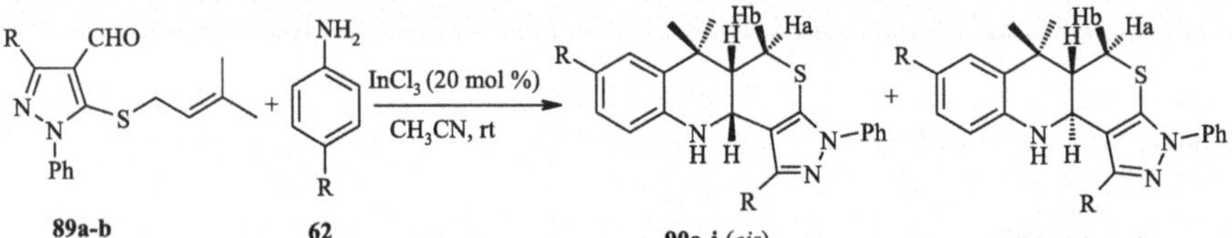

Scheme 37 Synthesis of pyrazolo-thiopyranoquinolines

In all the cases the *cis* isomers were the major products. The structures of the compounds were confirmed by spectral studies. The *cis* and *trans* stereochemistries of the products were assigned on the basis of the coupling constants of the protons in the ^{1}H NMR spectroscopy. All results are summarized in Table 20.

Similarly, three diastereomeric mixtures of bis4,4′-methylene or 4,4′-oxo-pyrazolo[4′,3′:5,6]-thiopyrano[4,3-*b*]quinolines **93a–d** (*cis-cis*), **94a–d** (*cis-trans*) and **95a–d** (*trans-trans*) have been prepared by the reaction of aromatic

Table 20 $InCl_3$ catalyzed synthesis of pyrazolo-thiopyranoquinolines

Entry	Aldehydes (89a–b) R	Anilines (62) R_1	Time (min)	Product ratio **90:91** (*cis:trans*)	Overall yield (%) (**90+91**)
a	CH_3	H	15	86:14	88
b	CH_3	OCH_3	20	90:10	92
c	CH_3	CH_3	25	9:91	90
d	CH_3	Cl	20	87:13	95
e	CH_3	NO_2	20	93:7	85
f	Ph	H	25	94:6	96
g	Ph	OCH_3	30	88:12	87
h	Ph	CH_3	20	90:10	90
i	Ph	Cl	15	92:8	92
j	Ph	NO_2	25	94:6	94

amines **62a–b** and *S*-allyl derivatives of pyrazoloaldehydes **89a–b** in the presence of 40 mol % $InCl_3$ via the intramolecular [4+2] imino-Diels–Alder cycloaddition (IMIDA) of imines (derived in situ by the reaction of anilines with the aldehyde functionality of pyrazolo aldehydes) with an olefinic moiety of *S*-allyl group in acetonitrile at room temperature in short reaction times with excellent overall yields (86–92%). All these isomers were separated by column chromatography. The product ratio was determined by the examination of the ^1H NMR spectrum of the crude product mixture. The stereochemistry of each isomer was assigned by ^1H NMR and nOe studies. In the *cis-trans* isomer **94a**, the coupling constant between H-3 and H-2 had a small J value (J_{2-3} = 2.6 Hz). This indicates *cis*-fusion at the ring junction, which was further confirmed by a strong nOe between H-3 and H-2. Also, the coupling constant between H-3′ and H-2′ had a large J value ($J_{2'-3'}$ = 11.0 Hz), which indicates a *trans*-ring fusion, which was further confirmed by the absence of an nOe. See Table 21 for the results (Scheme 38) [145].

Table 21 $InCl_3$ catalyzed synthesis of pyrazolo-thiopyranoquinolines

Entry	Aldehydes (89a–b) R	Anilines (92a–b) X	Time (min)	Product ratio **93:94:95** (*cis-cis:cis-trans: trans-trans*)	Overall yield (%)
a	CH_3	CH_2	35	85:10:5	92
b	CH_3	O	20	83:13:4	86
c	Ph	CH_2	25	84:10:6	90
d	Ph	O	20	86: 9:5	89

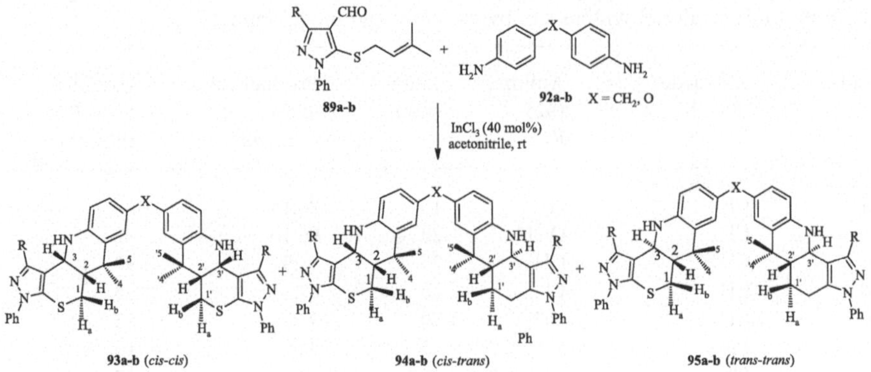

Scheme 38 Synthesis of pyrazolo-thiopyranoquinolines

5.2.13 Indolo-Pyrroloquinolines

Intramolecular imino-Diels–Alder reaction (IMIDA) of imines (derived in situ by the reaction of anilines **62** with the aldehyde functionality of indole **96**) with dienophile like the *N*-prenyl moiety of indole **96** in the presence of 20 mol % $InCl_3$ in acetonitrile at room temperature afforded the corresponding diastereomeric mixture of indolo[2,1-*a*]pyrrolo[4′,3′:2,3]-7a,8,13,13b-tetrahydroquinolines **97a–e** (*cis*) and **98a–e** (*trans*) in shorter reaction times with overall good yields (76–84%). In all the cases the *cis* isomers were the major isomers (Table 22 and Scheme 39) [146].

Among $AlCl_3$, BF_3cdotOEt_2, $ZnCl_2$ and $InCl_3$ used for this reaction, $InCl_3$ resulted in good yields and selectivities. The structures of the cycloadducts **97c** and **98c** were confirmed by the examination of their respective ^{1}H NMR spectra. The H_a proton of **97c** appears as a doublet at δ4.99 ($J = 6.3$ Hz), which was coupled to the H_b proton. This small coupling constant was consistent

Table 22 $InCl_3$ catalyzed synthesis of indolo-pyrroloquinolines

Entry	Anilines (62) R	Indole derivative	Time (min)	Product ratio *cis:trans* (97:98)	Overall yield (%) (97+98)
a	H	**96**	45	84:16	78
b	CH_3	**96**	30	80:20	84
c	OCH_3	**96**	35	88:12	80
d	Cl	**96**	30	92:8	76
e	NO_2	**96**	42	95:5	81

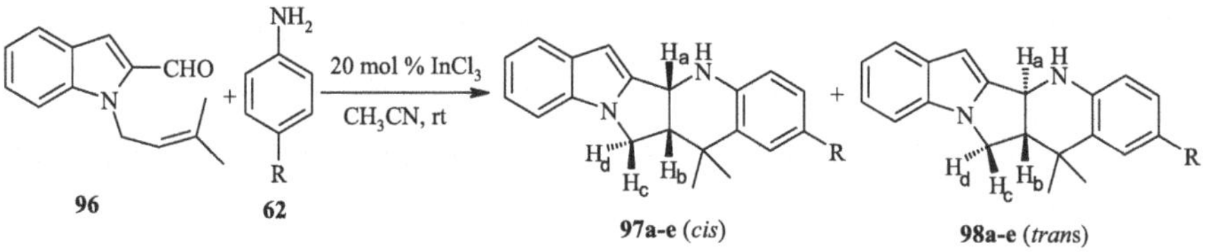

Scheme 39 Synthesis of indolo-pyrroloquinolines

with a *cis*-diaxial relationship for these two protons. Furthermore, the stereochemistry of H_a was confirmed by the observation of a strong (8.1%) nOe enhancement of H_a upon the irradiation of H_b. In a similar fashion, the H_a proton of **98c** exhibited a doublet at δ5.08 ($J = 10.6$ Hz) indicating the *trans* stereochemistry of H_a with respect to H_b.

Similarly, the [4+2] intramolecular imino-Diels–Alder reaction of imines (derived in situ by the reaction of anilines **62** with the aldehyde functionality of indole **99**) with dienophile as in the *N*-cinnamyl moiety of indole **99**, in the presence of 20 mol % $InCl_3$ in acetonitrile at room temperature provided the corresponding diastereomeric mixture of indolo-pyrroloquinolines **100a–e** (*cis*) and **101a–e** (*trans*) in short reaction times with excellent overall yields (86–92%). The structures of cycloadducts **100d** and **101d** were established by the examination of their ^{1}H NMR spectra and nOe experiments.

Table 23 $InCl_3$ catalyzed synthesis of indolo-pyrroloquinolines

Entry	Anilines (62) R	Indole derivative	Time (min)	Product ratio *cis:trans* (100:101)	Overall yield (%) (100+101)
a	H	**99**	45	87:13	86
b	CH_3	**99**	30	82:18	90
c	OCH_3	**99**	35	90:10	92
d	Cl	**99**	30	95:5	90
e	NO_2	**99**	42	96:4	87

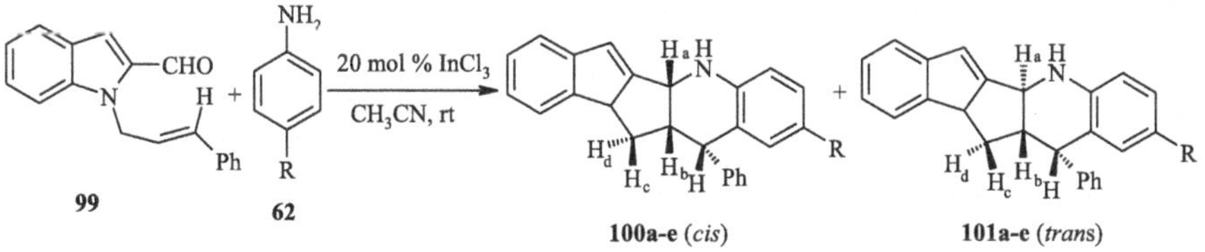

Scheme 40 Synthesis of indolo-pyrroloquinolines

The enhanced yields and short reaction times for the formation of products **100d** and **101d** relative to the cycloaddition of **96** with **62** (R = Cl) in acetonitrile is clearly a testament to the greater reactivity of the phenyl-substituted dienophile in **99**. The yields and ratios of the isomers are shown in Table 23 (Scheme 40) [146].

5.2.14
Quinoline-4-Hydrazones

The vinylogous aza-Povarov reaction between aromatic imines **1** and methacrolein dimethylhydrazone **102** in the presence of 10 mol % $InCl_3$ furnished C-4 functionalized 1,2,3,4-tetrahydroquinolines **103a–h** bearing a hydrazone function at C-4 in good to excellent yields (Scheme 41 and Table 24). In this reaction, the imine acts as the diene and due to the electron-withdrawing effect of the nitrogen atom the C=C portion of the α,β-unsaturated dimethylhydrazone acts as the dienophile. This reaction tolerates well the presence of both electron-withdrawing and electron-releasing groups at both aromatic rings of the starting imine. Additional advantage of this reaction involves the creation of two C–C bonds and the generation of two stereocenters, one of

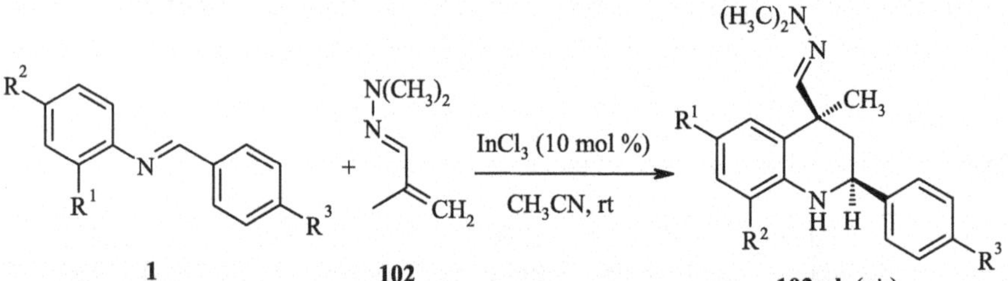

Scheme 41 Synthesis of quinoline-4-hydrazones

Table 24 $InCl_3$ catalyzed synthesis of quinoline-4-hydrazones

Schiff bases (1) R¹	R²	R³	Time (h)	Product	Yield (%)
H	OCH_3	H	2	**103a**	90
H	CH_3	Cl	2	**103b**	76
H	OCH_3	Cl	3	**103c**	87
H	OCH_3	CH_3	2	**103d**	93
OCH_3	H	Cl	3	**103e**	70
CH_3	CH_3	Cl	3	**103f**	79
CH_3	CH_3	OCH_3	3	**103g**	83
CH_3	CH_3	Br	3	**103h**	80

them quaternary, with complete diastereoselectivity and in a single synthetic operation [147].

The vinylogous aza-Povarov reaction was stereoselective and afforded exclusively compounds **103** with a *cis* relationship between the C-2 hydrogen and the C-4 methyl. This structure was established by NOESY studies, which showed a correlation peak between the H-2 and C_4-CH_3 signals, which was only consistent with a *cis* arrangement for these substituents, both being equatorial. The observed stereoselectivity could be explained by taking into account that the Povarov reaction was known to proceed in a stepwise manner, as proved by the trapping of the intermediate iminium species when the reaction was carried out in the presence of nucleophiles [148]. Therefore, the generation of intermediate **104** from the reaction between dimethylhydrazone **102** and indium–imine complex **105** was expected to form under these reaction conditions. The final cyclization step takes place through

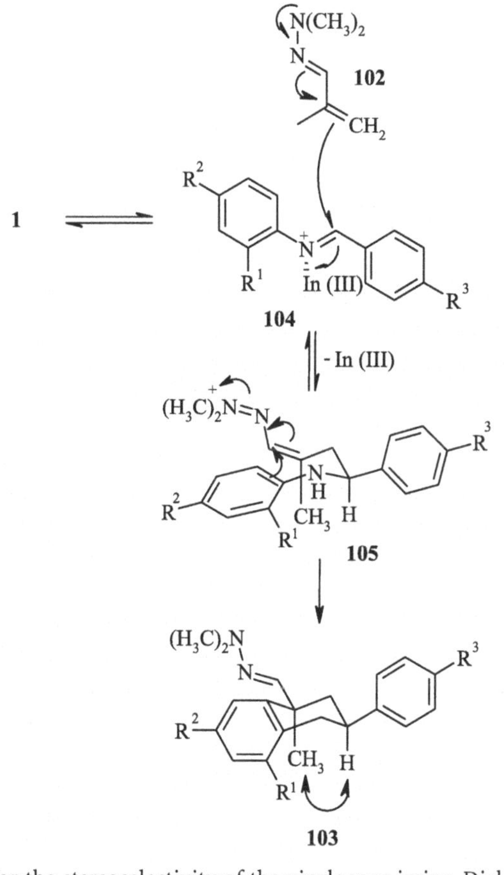

Fig. 14 Rationale for the stereoselectivity of the vinylogous imino-Diels–Alder reaction

a chair-like transition state, leading to a preference for an equatorial arrangement of the bulky dimethylhydrazono and aryl substituents and hence to the observed stereochemistry for compounds **103** (Fig. 14). The generation with complete control of the stereocenter at the tetrahydroquinoline C-4 position is noteworthy, since the stereoselective installation of quaternary stereocenters is one of the most challenging tasks for synthetic organic chemistry [149].

5.2.15 Quinoline-4-Carboxylic Acids

Quinoline-4-carboxylic esters or amides **107a–m** have been synthesized by the reaction of 2-methoxyacrylates or acrylamides **106a–d** with *N*-arylbenzaldimines **1** in acetonitrile under 50 mol % $InCl_3$ catalysis and a 150 W microwave irradiation for 3 min in moderate yields (20–57%) (Scheme 42) [150].

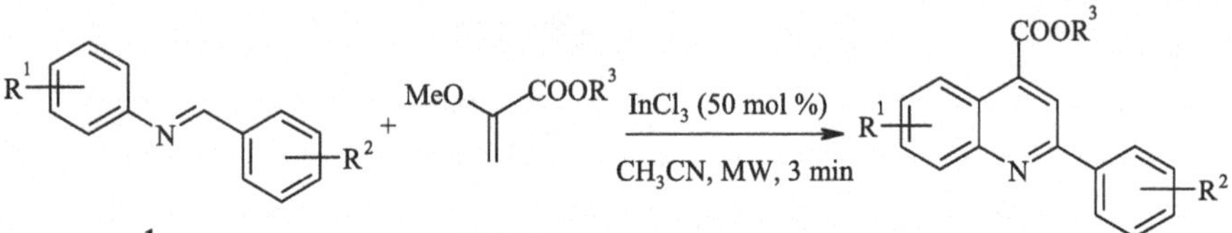

Scheme 42 Synthesis of quinoline-4-carboxylic acids

In this reaction, the important step was the imino-Diels–Alder reaction of *N*-arylbenzaldimines as dienes and 2-methoxyacrylates or amides as dienophiles to get tetrahydroquinolines, which further converted to quinolines rapidly. This reaction did not yield satisfactory results with other Lewis acids such as $BiCl_3$, $ZnCl_2$, $SnCl_4$, InF_3, $In(O^tBu)_3$, $Sc(OTf)_3$, $InBr_3$, InI_3. The Lewis acid and the microwave activation appeared as crucial parameters for obtaining the reaction in satisfactory yields and in determining the reactivity of the partners in the formal imino-Diels–Alder reaction. The scope and the generality of this procedure are presented in Table 25. The cycloaddition appears to be dependent on the substituents on both aromatic rings. On the aldehydic end of the imine an electron-withdrawing group in the *ortho* or *meta*-position impeded the cycloaddition (**107c,e**) whereas, a halogen in these positions allowed the formation of quinolines (**107b,d,j,l**) in 35–43% yields. No such effect resulted from substitution in the *para*-position (**107g,h**). When the aniline component was substituted with a halogen at the *para*-position, the best yields were observed with less electron-withdrawing atoms (**107f,i**). Finally, comparisons of the yields obtained in the formation of fluoroquinoline carboxylic esters and amides showed that higher yields were obtained with amides wherever the halogen was located.

Table 25 $InCl_3$ catalyzed synthesis of quinoline-4-carboxylic acids

N-Arylbenzaldimines 1			Product	Yield (%)
R^1	R^2	R^3		
OR	**107a**	H	**107a**	57
OR	**107b**	2′-F	**107b**	40
OR	**107c**	2-CF_3	**107c**	0
OR	**107d**	3′-Br	**107d**	45
OR	**107e**	3′-NO_2	**107e**	0
OR	**107f**	H	**107f**	24
OR	**107g**	4′-CN	**107g**	32
OR	**107h**	4′-NO_2	**107h**	23
OR	**107i**	H	**107i**	52
NH^iPr	**107j**	2′-F	**107j**	43
NHCH(Ph)Et	**107k**	2′-F	**107k**	35
NHCH(Ph)Et	**107l**	4′-F	**107l**	48
NHCH(Ph)Et	**107m**	H	**107m**	49

5.2.16
Benzonaphthyridines and Indoloquinolizidines

Benzonaphthyridine derivatives exhibit a wide spectrum of biological activity such as bactericidal, fungicidal, and cancerostatic [151–153]. Also they are interesting ligands of the Werner-type σ-complexes with metal central atoms as well as EDA π-complexes [154].

Diastereomeric benzo[*h*][1,6]naphthyridines **110a–b** (*trans*) and **111a–b** (*cis*) were synthesized by a three-component reaction of dihydropyridines **108a–b**, *p*-methylaniline **62b** and ethylglyoxalate **109** in the presence of 20 mol % $InCl_3$ via [4+2] imino-Diels–Alder cycloaddition between the olefinic moiety of dihydropyridine with the imine moiety (derived in situ by

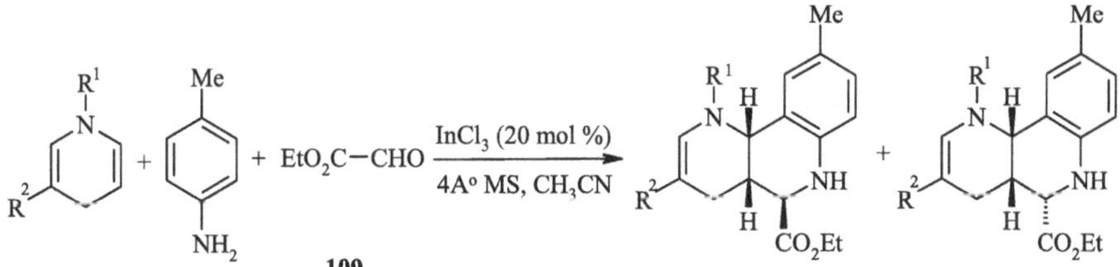

108a-b **62b** **109** **110a-b** (*trans*) **111a-b** (*cis*)

108a: R^1 = Me, R^2 = CN

108b: R^1 = Me, R^2 = CO_2Me

Scheme 43 Synthesis of benzonaphthyridines

the reaction of *p*-methylaniline and ethylglyoxalate) in acetonitrile at room temperature with moderate overall yields (63–65%). This reaction works well with different substituents (Me, CN, CO_2Me) attached to the dihydropyridine (Scheme 43) [155].

Similarly the reaction of *N*-methoxycarbonyl-1,2-dihydropyridine **112** with *p*-methylaniline **62b** and ethyl glyoxalate **109** afforded the corresponding benzonaphthyridine **113** as a single isomer (*cis*) under the same reaction conditions in 37% yield (Scheme 44) [155].

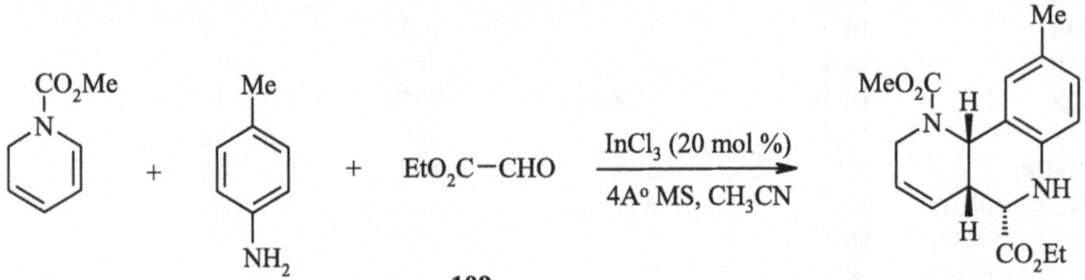

Scheme 44 Synthesis of benzonaphthyridines

The reaction of dihydroisoquinoline **114** with *p*-methylaniline **62b** and 4-chlorobenzaldehyde **115** under reflux conditions in acetonitrile for 48 h furnished the desired dibenzo[*c*, *h*][1,6]naphthyridines **116** (*cis*) and **117** (*trans*) in 25% overall yield (Scheme 45) [155].

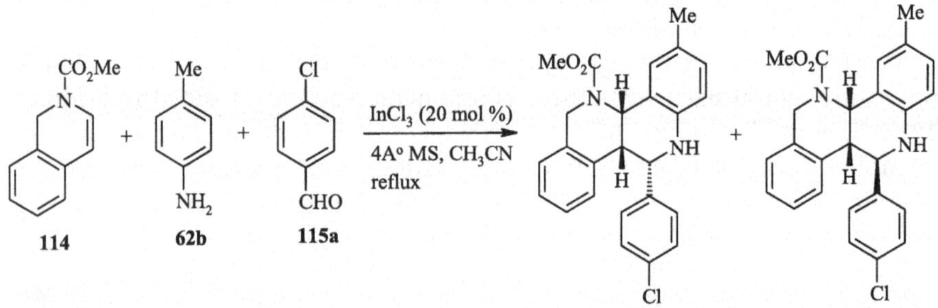

Scheme 45 Synthesis of benzonaphthyridines

The scope of this method with regard to further modifications on the aldehyde and aniline components was investigated. This reaction was not successful when paraformaldehyde, glyoxylic acid and chloral hydrate were used instead of ethyl glyoxalate **109**. On the other hand, the reaction with highly activated 3-methoxyaniline occurred through aniline addition upon the intermediate imine [156], by-passing the interaction with the dihydropyridine **108**.

However, the reaction of the electron-deficient amine ethyl *p*-aminobenzoate **62c** reacted with dihydropyridine **118** (generated by cyanide addition upon the corresponding *N*-tryptophylpyridinium salt) proceeded with ethyl glyoxylate **109** under $InCl_3$ catalysis to afford stereoselectively the indolo[3,2-*a*]quinolizidine derivative **119** in 66% overall yield (as a 4:1 mixture of epimers at the α-aminoester center, stereochemistry not determined) (Scheme 46).

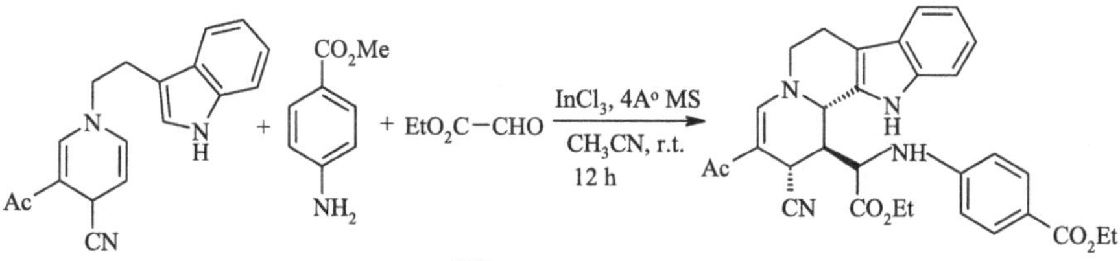

118 **62c** **109** **119**

Scheme 46 Synthesis of indoloquinolizidine

The stereochemical outcome can be rationalized by considering the preferential attack of the dihydropyridine from its less hindered face and the final indole cyclization upon the iminium ion taking place in a stereocontrolled manner to yield a *trans*-indolo[3,2-*a*]quinolizidine derivatives [155].

5.2.17
Pyridine and Quinoline-β-Lactams and Indolizidines

Indolizidine alkaloids have recently attracted a lot of attention due to their widespread occurrence and their utility as research tools in pharmacology. Their structural and stereochemical complexity, coupled with their diverse and potent biological activities, makes indolizidine alkaloids as well as related non-natural compounds very attractive synthetic targets in the search for efficient and selective synthetic methods. In addition, functionalized bicyclic lactams structurally related to indolizidine have been found to act as conformationally restricted peptide mimetics, and they have been utilized in asymmetric synthesis leading to natural products [157–163].

Imines derived from 4-oxoazetidine-2-carbaldehydes have been found to be versatile Diels–Alder reagents in that they exhibit two reactivity patterns. These imines lead to cycloadducts arising from normal as well as inverse electron-demand [4+2] cycloaddition. Accordingly, the aza-Diels–Alder reaction of aldimine **120a–c** with Danishefsky diene **121** in the presence of 20 mol % $InCl_3$ in aceteonitrile at −20 °C afforded the corresponding diastereomeric azetidine-dihydropyridin-4-ones **122a–c** (*anti*) and **123a–c** (*syn*) in good overall yields favoring *anti* stereoselectivity; results are shown in Table 26. In this aza-Diels–Alder reaction 2-azetidinone-tethered imines

Table 26 $InCl_3$ catalyzed synthesis of tetrahydropyridin-4-ones

Entry	Imine	R^1	R^2	R^3	Product ratio 122:123	Overall yield (%) (122+123)
a	**120a**	PMP	PhO	PMP	78:22	81
b	**120b**	Tol	MeO	PMP	75:25	57
c	**120c**	PMP	Ph	Bn	100:0	64

120 act as dienophiles. These diastereomers were easily separated by gravity flow chromatography (Scheme 47).

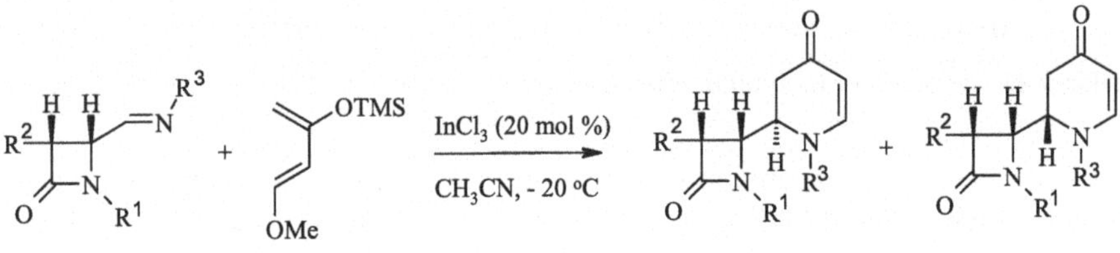

120a-c **121** **122a-c** (*anti*) **123a-c** (*syn*)

Scheme 47 Synthesis of tetrahydropyridin-4-ones

Other Lewis acids like $In(OTf)_3$ and ZnI_2 also furnished good results. These diastereomers **122b** or **123b** have been converted to the corresponding highly functionalized indolizidines **124** or **125** via multi-step functional group transformations (Scheme 48) [164].

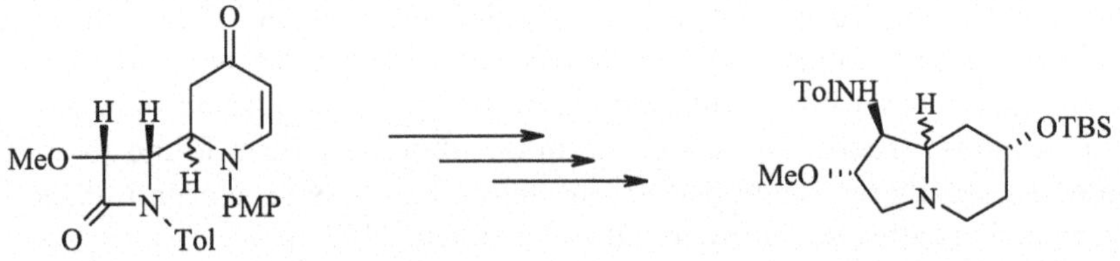

122b or 123b **124 or 125**

Scheme 48 Synthesis of indolizidines

The vicinal coupling constants of the two protons (H_4 in the β-lactamic ring, hydrogen α to the nitrogen in the six-membered ring) were diagnostic of the relative stereochemistry of these stereocenters. The vicinal coupling constant for cycloadducts **122** is approximately 10.0 Hz, which suggests a relative *anti* stereochemistry for this connection, whereas these vicinal coupling constant for the minor isomers **123** was approximately 3.0 Hz, and indicate a relative *syn* stereochemistry. This configurational assignment was further confirmed by means of an X-ray diffraction analysis of these cycloadducts. The stereochemical course of the reaction could be explained by assuming that the Lewis acid catalyst coordinates to the imine nitrogen and that the β-lactam moiety preferentially adopts an anti-Felkin–Anh conformation, in which the large substituent of the four-membered ring (the amino group) was oriented perpendicular to the imine group. The silyloxydiene should then preferentially attack from the backside to furnish the major product (Fig. 15) [165].

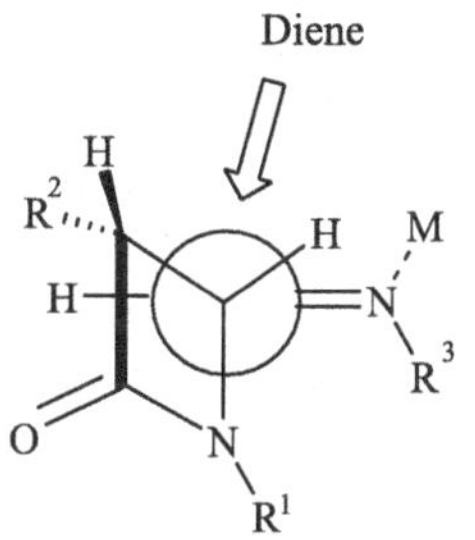

Fig. 15 Model showing the origin of the observed *syn* stereochemistry

The stereoselectivity of the aza-Diels–Alder reaction was found to be dependent on the bulkiness of the *N*-substituents on the imines. When less hindered *N*-benzyl imine was used other than *N*-4-methoxyphenyl imine for cycloadduct formation, fully diastereoselective conversion was obtained (Table 26, **120c**).

Aza-Diels–Alder reaction of aldimines **120a–e** with less electron-rich diene 2,3-dimethyl-1,3-butadiene **126** in the presence of 20 mol % $In(OTf)_3$ in aceteonitrile at 0 °C afforded the corresponding diastereomeric cycloadducts

Table 27 $In(OTf)_3$ catalyzed synthesis of quinoline-β-lactams

Entry	Imine	R^1	R^2	Product ratio **127**:**128**	Overall yield (%) (**127**+**128**)
a	**120a**	PMP	PhO	60:40	85
b	**120b**	Tol	MeO	75:25	83
c	**120c**	PMP	Ph	65:35	86
d	**120d**	PMP	MeO	65:35	89
e	**120e**	Allyl	MeO	100:0	67

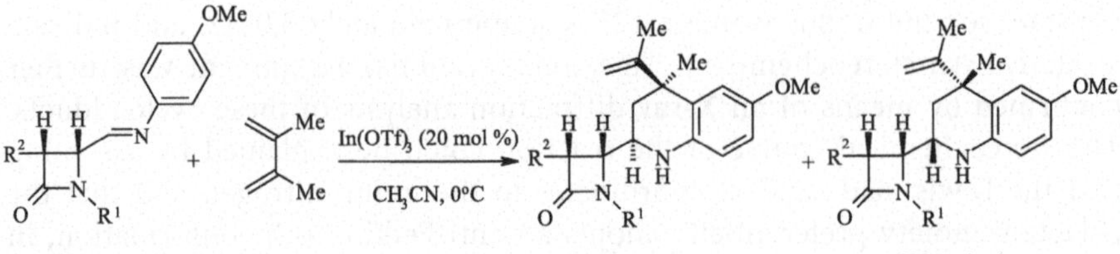

120a–e 126 127a–e (*anti*) 128a–e (*syn*)

Scheme 49 Synthesis of quinoline-β-lactams

127a–e (*anti*) and **128a–e** (*syn*) in good overall yields favoring *anti* stereoselectivity and the results are shown in Table 27 (Scheme 49).

Interestingly, the dienophilic behavior of imines **120** in the Diels–Alder reaction was reversed, such that it exhibited heterodienic properties [166–168]. The cycloaddition did not take place with electron-poor dienophiles such as 2-cyclohexen-1-one or methyl acrylate, confirming that an inverse electron-demand Diels–Alder reaction was involved. This azadiene behavior was well known for aryl imines derived from aromatic or α,β-unsaturated aldehydes [169, 170], but little was known about the use of aliphatic aldehyde derived-imines as the 4π component [171, 172] and even less about their optically active derivatives [173]. Furthermore, apparently the dual behavior of aliphatic aldehyde-derived imines as both azadienes and as dienophiles was unprecedented. These diastereomers **127a** or **128a** were easily converted to the corresponding indolizidines **129** or **130** via multi-step functional group transformations (Scheme 50).

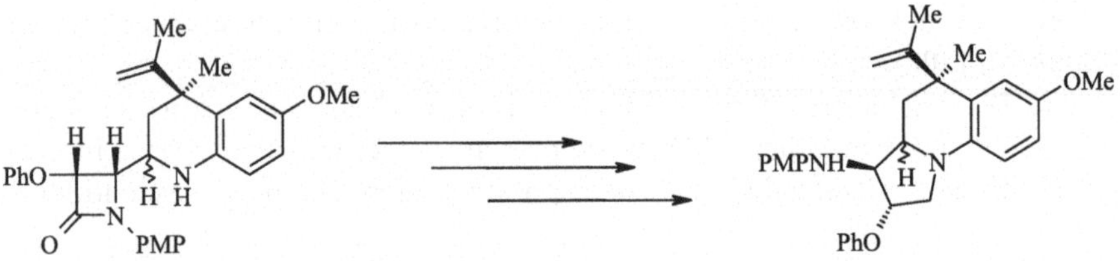

127a or 128a 129 or 130

Scheme 50 Synthesis of indolizidines

Similarly, aza-Diels–Alder reaction of aldimine **120d** with a cyclic alkene like cyclopentadiene **43** in the presence of 20 mol % In(OTf)$_3$ in acetonitrile at –20 °C afforded the corresponding diastereomeric cyclopentaquinoline-β-lactams **131** (*anti*) and **132** (*syn*) in excellent overall yields (98%) (Scheme 51), which were then easily converted to corresponding indolizidines **133** or **134** via multi-step functional group transformations (Scheme 52).

In the same manner, aza-Diels–Alder reaction of aldimine **120a,d,e** with 3,4-dihydro-2*H*-pyran **55** in the presence of 20 mol % In(OTf)$_3$ in ace-

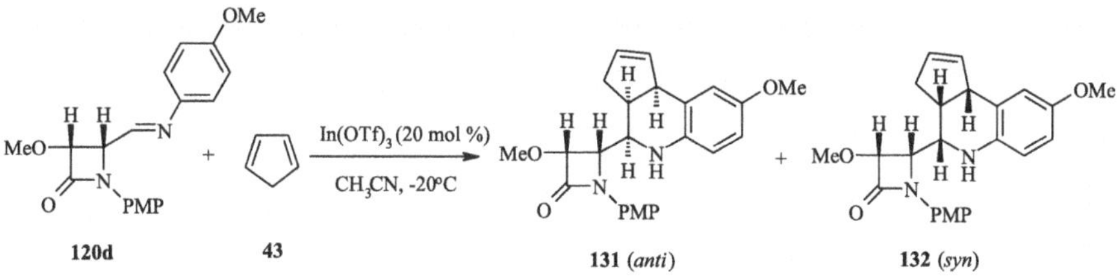

Scheme 51 Synthesis of cyclopentaquinoline-β-lactams

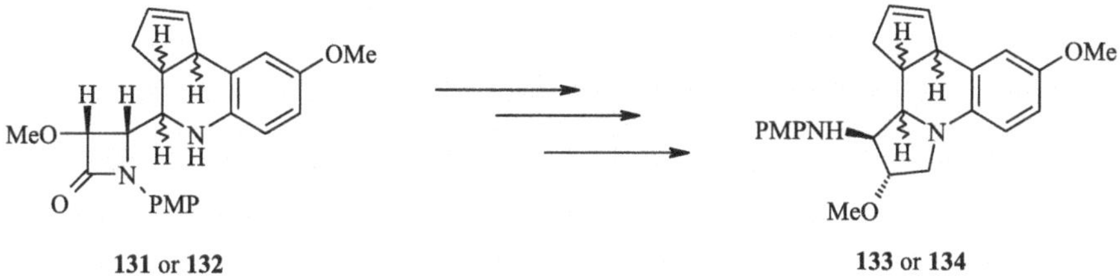

Scheme 52 Synthesis of indolizidines

teonitrile at $-20\,^{\circ}$C provided the corresponding diastereomeric pyrano[3,2-*c*]quinoline-β-lactams **135a,d,e** (*anti*) and **136a,d** (*syn*) in good overall yields (Scheme 53).

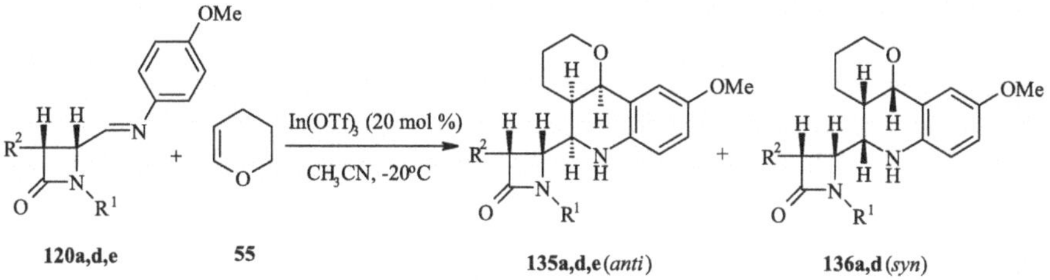

Scheme 53 Synthesis of pyrano[3,2-*c*]quinoline-β-lactams

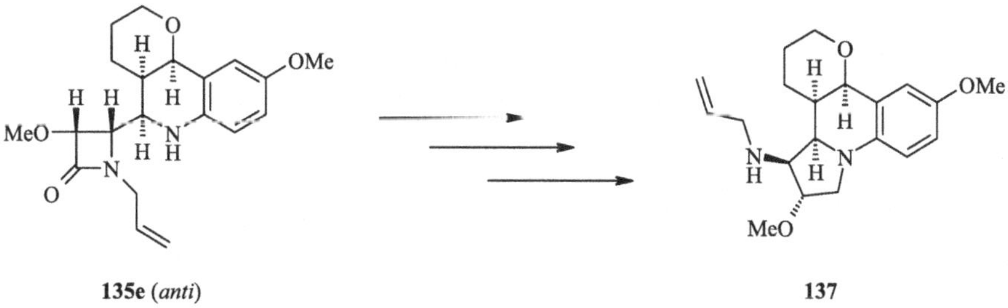

Scheme 54 Synthesis of indolizidines

Furthermore, pyrano[3,2-*c*]quinoline-β-lactam **135e** (*anti*) was easily converted to the corresponding indolizidines **137** via multi-step functional group transformations (Scheme 54).

5.2.18 Piperidines

Functionalized piperidines occur with great regularity in the natural product arena and as important units in pharmaceuticals. Over the last ten years thousands of piperidine-containing compounds have been entered into preclinical and clinical trials [174].

Highly functionalized piperidines **139a–i** have been prepared by the five-component condensation which involves the reaction of 1 eq of methyl acetoacetate **138**, 2 eq of aldehydes **115** and 2 eq of anilines **62** in a one-pot, atom and step economic (PASE) process in the presence of 33 mol % $InCl_3$ in acetonitrile at room temperature in 16–74% overall yield (Scheme 55) [175].

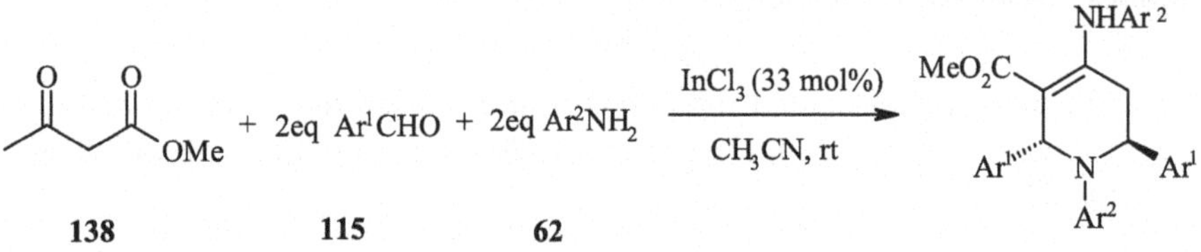

Scheme 55 Synthesis of piperidines

This reaction was highly diastereoselective with respect to 2,6-diphenyl groups, whose relative stereochemistry was confirmed as *trans* by single crystal X-ray analysis. In most cases, the piperidine precipitates out of the re-

Table 28 $InCl_3$ catalyzed synthesis of piperidines

Entry	Ar^1CHO (**115**)	Ar^2NH_2 (**62**)	Product yield (%) (**139**)
a	4-$CH_3C_6H_4$	4-$CH_3OC_6H_4$	45
b	3-$CH_3C_6H_4$	4-$CH_3OC_6H_4$	48
c	Ph	4-$CH_3OC_6H_4$	74
d	3-$CF_3C_6H_4$	4-$CH_3OC_6H_4$	57
e	4-CH_3OC_6H4	Ph	52
f	3-$CH_3C_6H_4$	Ph	64
g	Ph	Ph	60
h	4-$NO_2C_6H_4$	Ph	52
i	2-$CH_3C_6H_4$	Ph	16

action mixture allowing for easy isolation. The scope and generality of this reaction are illustrated in Table 28.

The electron-donating group attached to anilines increases the rate of reaction compared to an electron-withdrawing group. Also yields were found to be lower when two-substituted aldehydes were employed, probably due to steric effects. This reaction failed to yield piperidines when aliphatic aldehydes were employed. The plausible mechanism for the formation of piperidine **139g** involves the Lewis acid promoted formation of enamine **140**, which then undergoes a "Knoevenagel-like" condensation with benzaldehyde **115a** to form the iminium ion–Knoevenagel product **141**. Loss of a proton and tautomerization of the imine to the enamine generates a diene **142**, which then undergoes either an aza-Diels–Alder cyclization or a tandem Mannich–Michael reaction with imine **1** to furnish piperidine **139g** (Fig. 16).

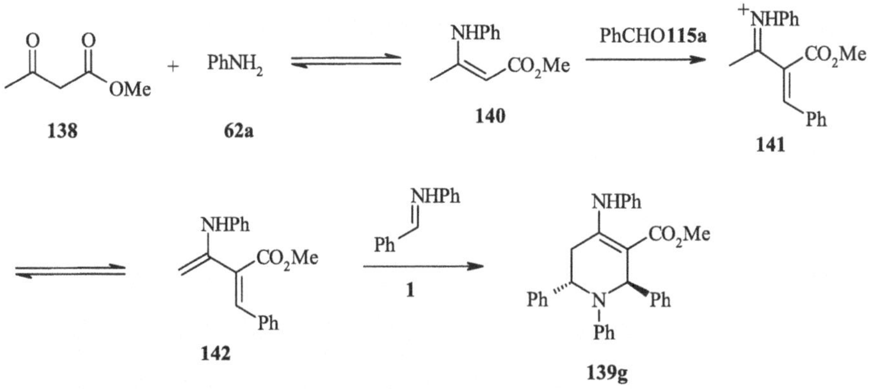

Fig. 16 The plausible mechanism for the formation of piperidine **139g**

5.2.19 Pyridin-4-Ones

The imino-Diels–Alder reaction of Schiff bases **1** (generated in situ by the reaction of amines **62** with aromatic aldehydes **115**) with 1-methoxy-3-trimethylsilyloxy-1,3-butadiene (Danishefsky's diene) **121** in the presence of 0.5 mol% $In(OTf)_3$ resulted in corresponding pyridine-4-ones **143a–f** in

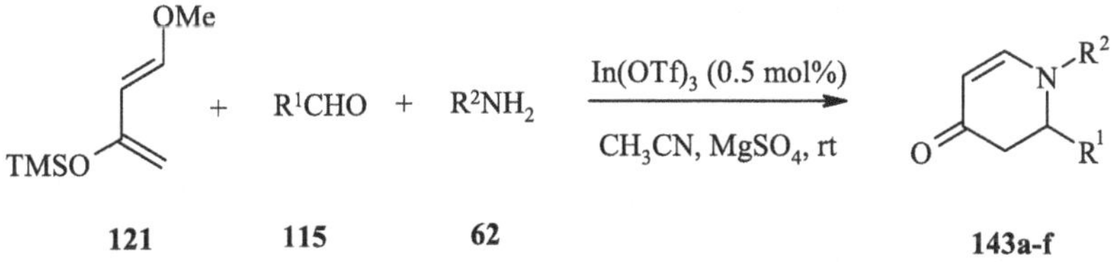

Scheme 56 Synthesis of pyridine-4-ones

Table 29 $InCl_3$ catalyzed synthesis of pyridine-4-ones

Aldehydes (**115**) R^1	Anilines (**62**) R^2	Product	Yield (%)
C_6H_5	C_6H_5	**143a**	51
2-FC_6H_4	C_6H_5	**143b**	84
2-FC_6H_4	$C_6H_5CH_2$	**143c**	50
2-Pyridyl	C_6H_5	**143d**	95
2-Furyl	C_6H_5	**143e**	61
2-Thiphenyl	C_6H_5	**143f**	0

moderate to good yields (50–95%) within short reaction time (30 min) at room temperature. $Sc(OTf)_3$ also provided the same results. However, the Schiff base derived from 2-thiophene carboxaldehyde did not yield any adduct **143f** and the results are summarized in Table 29 (Scheme 56) [176].

5.2.20 Azabicyclooctanones and Azabicyclononanones

N-Benzylidene anilines **1** reacted with cyclohexen-2-one **144** in the presence of 20 mol % $InCl_3$ to afford the corresponding diastereomeric azabicyclo[2.2.2]octanone derivatives (**145:146a–e**) via imino-Diels–Alder reaction, in contrast to the formation of expected phenanthridinone derivatives (**147:148a–e**). All the reactions proceeded smoothly under mild reaction conditions to afford the azabicyclo[2.2.2]octanones in moderate yields (60–74%) [105] and the results were summarized in Table 30 (Scheme 57).

The reaction seems to proceed through the formation of dienolate ion by strong coordination of $InCl_3$ with cyclohexen-2-one **144**. Thus, the resulting cyclohexadienolate ion, which acts as the diene underwent imino-Diels–Alder reaction with the Schiff base, which acts as the dienophile. Alternatively, the formation of azabicyclooctanones (**145:146a–e**) were also

Table 30 $InCl_3$ catalyzed synthesis of azabicyclo[2.2.2]octanones

Entry	Schiff bases (1) R	Ar	Product ratio (**145:146**)	Overall yield (**145+146**) (%)
a	H	C_6H_5	69:31	65
b	NO_2	C_6H_5	67:33	70
c	OCH_3	C_6H_5	68:32	60
d	Cl	C_6H_5	73:27	68
e	Cl	*p*-Cl–C_6H_4	47:53	74

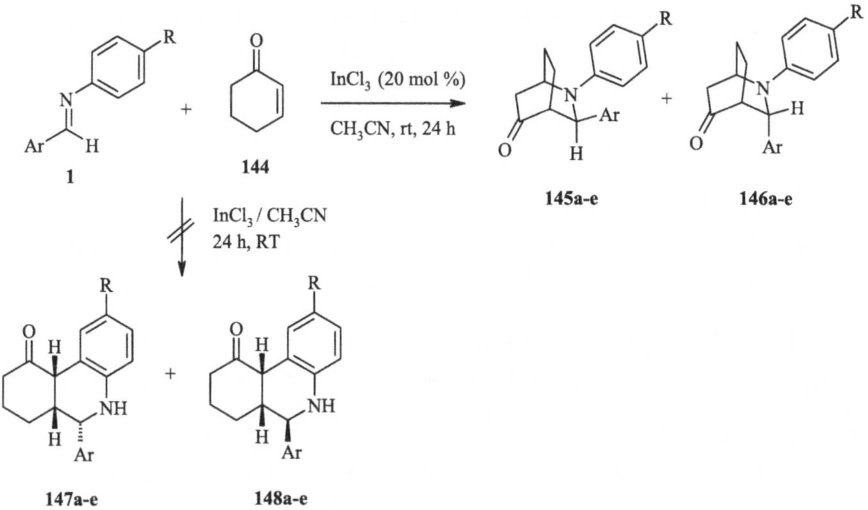

Scheme 57 Synthesis of azabicyclooctanones

Table 31 InCl$_3$ catalyzed synthesis of azabicyclo[2.2.3]nonanones

Entry	Schiff bases (1) R	Ar	Product ratio (**150**:**151**)	Time (h)	Overall yield (%) (**145**+**146**)
a	H	C_6H_5	52:48	24	56
b	OCH_3	C_6H_5	65:35	26	48
c	Cl	C_6H_5	39:61	20	78
d	Cl	*p*-Cl-C_6H_5	42:58	21	68
e	CH_3	C_6H_5	48:52	22	52

accounted by Mukiyama aldol reaction of a diene, generated in situ from cyclohexene-2-one with the Schiff base followed by cyclization under Michael conditions [177, 178]. The structure of the product was ascertained by single crystal X-ray analysis of the adduct **145d** [179]. Similarly, imino-Diels–Alder

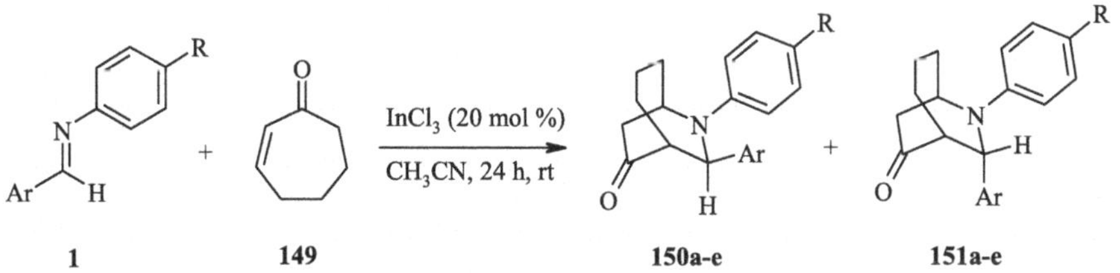

Scheme 58 Synthesis of azabicyclononanones

reaction of *N*-benzylidene anilines **1** with cycloheptene-2-one **149** afforded diastereomeric azabicyclo[2.2.3]nonanones (**150:151a–e**) in moderate yields (48–78%) and the results were shown in Table 31 (Scheme 58) [105].

5.3 Oxa-Diels–Alder Reactions

Oxa-Diels–Alder reactions basically adopt two variants (Fig. 17) [180]. In the first one, carbonyl compounds undergo [π2+π4] cycloaddition to afford 3,4-dihydro-2*H*-pyrans. Dihydropyrans are important substrates for the synthesis of carbohydrates and other natural products [181]. Carbonyl compounds are in general of limited reactivity in oxa-Diels–Alder reactions. Only electron-deficient carbonyl compounds such as glyoxylates, chloral, ketomalonate, 1,2,3-triketones and related compounds react with dienes having electron-donating groups. However, electron-rich carbonyls require the use of Lewis acids or high pressures in order to undergo oxa-Diels–Alder reactions [182, 183].

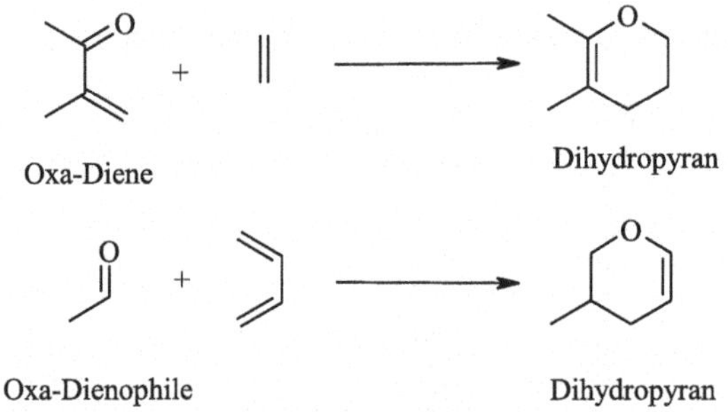

Fig. 17 Basic variants in oxa-Diels–Alder reactions

The α,β-unsaturated carbonyl compounds can be utilized in an inverse-electron demand Diels–Alder reaction as an electron-poor diene component, with suitably substituted alkenes to yield 3,4-dihydro-2*H*-pyrans [184].

5.3.1 Xanthenes

Xanthene derivatives are parent compounds of a large number of naturally occurring, as well as synthetic derivatives and occupy a prominent position in medicinal chemistry [185–187]. Especially, 4-methoxyxanthene-9-amine derivatives have been used for the synthesis of pharmacologically important Clavizepine alkaloids [188]. Cytotoxic natural stilbenes schweinfurthin A & B and vedelianin were found to possess a common left-half hexahydroxanthene ring system (Fig. 18) [189].

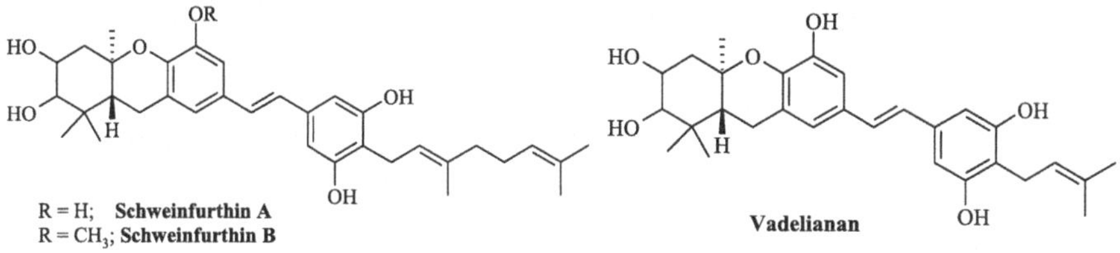

Fig. 18 Biologically important xanthene natural products

Novel xanthene-9-*N*-arylamine derivatives **154a–j** have been synthesized by the reaction of 1-(4-morpholino)-cyclohexene **152** with salicylaldehyde imines **153a–j** in the presence of 20 mol % $InCl_3$ in CH_3CN at room temperature in excellent yields within a short time (20–40 min). In all these reactions, a single product **154** was obtained, which upon recrystallization with ethyl acetate yielded pure crystalline products. The structures of the products were assigned by ^{1}H, ^{13}C NMR and X-ray crystallography studies. The results were summarized in Table 32 (Scheme 59) [190].

Table 32 $InCl_3$ catalyzed synthesis of xanthenes

R^1	R^2	Time (min)	Product	Yield (%)
C_6H_5	H	20	**154a**	96
2-CH_3-C_6H_4	H	30	**154b**	89
4-CH_3O-C_6H_4	H	20	**154c**	93
4-Br-C_6H_4	H	40	**154d**	90
2-Pyridine	H	30	**154e**	87
C_6H_5	5-CHO	30	**154f**	95
4-NO_2-C_6H_4	H	35	**154g**	82
4-CH_3-C_6H_4	7-Cl	20	**154h**	90
2-Pyridine	5-CH_3O	40	**154i**	86
2-Napthyl	5-CH_3O	20	**154j**	91

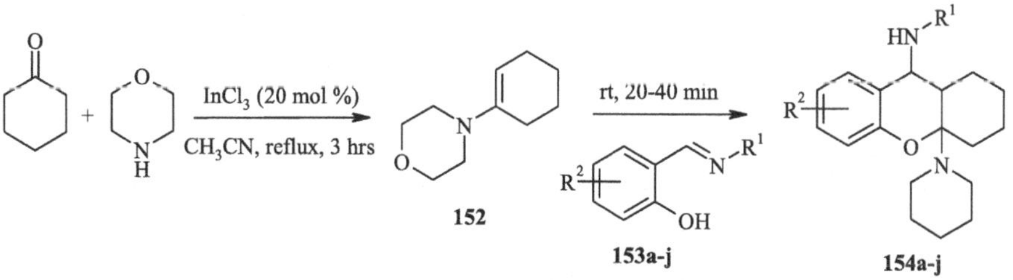

Scheme 59 Synthesis of xanthenes

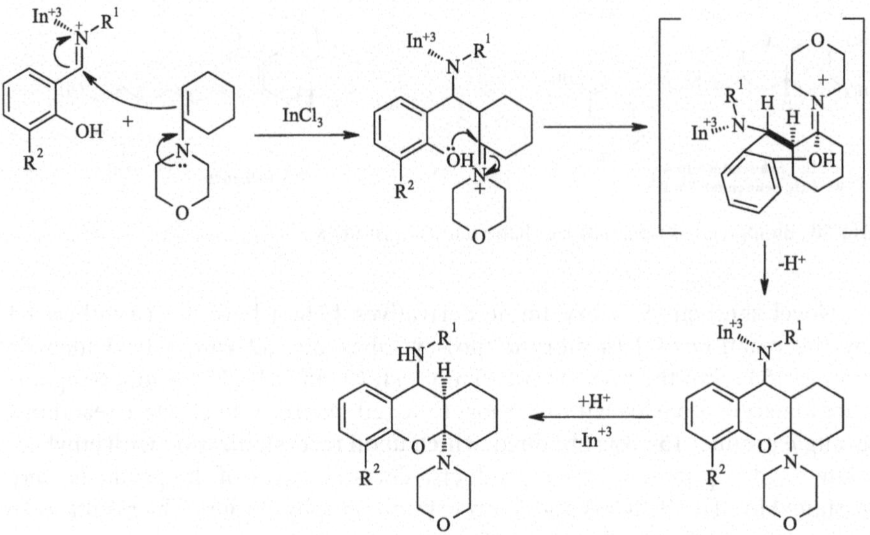

Fig. 19 Mechanism for the $InCl_3$ catalyzed synthesis of hexahydroxanthene-9-*N*-arylamines

The reaction is expected to proceed through the activation of the imine nitrogen by co-ordination of the catalyst $InCl_3$, followed by nucleophilic addition of the enamine to the C=N bond and subsequent cyclization of the iminium ion, resulting in the formation of the linear-fused hexahydroxanthene-9-*N*-arylamines (Fig. 19).

5.3.2 Dihydropyran-2-Ones (δ-Lactones)

δ-Lactones are important intermediates in the synthesis of a variety of structures with biological and medical significance, such as antifungal and antitumor activity [191–193]. The first novel efficient approach for the synthesis of 5,6-dihydro-6,6-disubstituted pyran-2-ones **157a–j** was reported from the reaction between commercially available ethyl 3-ethoxybut-2-enoate **155** (intermediate of Brassard's diene) and ketones **156a–j** in the presence of $BF_3{\cdot}Et_2O$

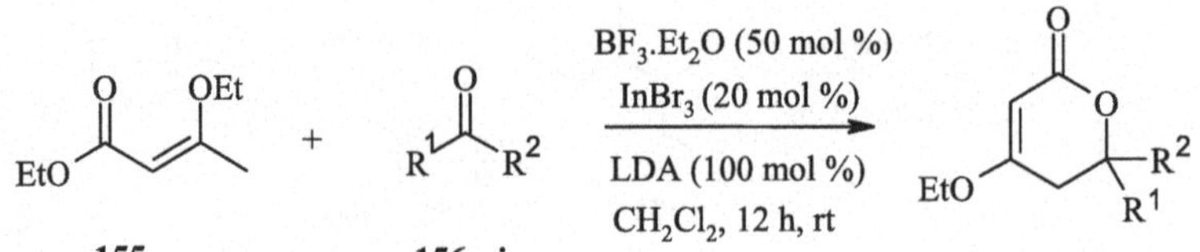

155 **156a-j** **157a-j**

Scheme 60 Synthesis of pyran-2-ones

(50 mol %), $InBr_3$ (20 mol %) and LDA (100 mol %) at room temperature in the reaction time of 12 h (Scheme 60) [194].

In this reaction, $BF_3 \cdot Et_2O$ and $InBr_3$ were employed as the Lewis acid to activate the ketone and LDA was employed as the Lewis base to activate ethyl 3-ethoxybut-2-enoate **155**. The other Lewis acids $TiCl_4$, $ZnCl_2$, $Yb(OTf)_3$, $Sc(OTf)_3$, Et_2AlCl do not yield any products. Various aromatic, aliphatic, and heterocyclic ketones were converted to the corresponding δ-lactones containing quaternary carbon atom centers in excellent yields (63–99%) under mild conditions (see Table 33).

Table 33 $InBr_3$ catalyzed synthesis of pyran-2-ones

R^1	R^2	Product	Yield (%)
C_6H_5	CH_3	**157a**	99
C_6H_5	2-F–C_6H_4	**157b**	72
4-Cl–C_6H_4	CH_3	**157c**	68
4-CH_3–C_6H_4	CH_3	**157d**	99
4-OCH_3–C_6H_4	CH_3	**157e**	98
4-NO_2–C_6H_4	CH_3	**157f**	82
Cyclohexyl	CH_3	**157g**	81
tert-butyl	CH_3	**157h**	65
thiophenyl	CH_3	**157i**	63
C_6H_5	OEt–CH_2–OEt	**157j**	65

The efficiency of this method was associated with the electronic and steric effects of the substituents attached to ketones. With the enhancement of the substituent's electron-donating capability in the starting aromatic ketone, e.g., CH_3, OCH_3, the yield of pyranone increased (**157c,d,e**). Interestingly, also groups with electron-withdrawing capability, like Cl and NO_2, led to good yields (**157f**). This indicated that appropriate electronegative groups on substituted ketones could give high yields. The bulkier *ortho*-group, like isopropyl and *tert*-butyl, provided only moderate yields (**157h**), which could be attributed to a steric effect. Also aromatic ketones were superior to aliphatic ketones. Moreover, it is worthwhile to note that a heterocyclic ketone was transformed with moderate yield to pyranone (**157i**). Surprisingly acetal-substituted active ketone was transformed to the corresponding pyranone (**157j**) in 65% yield.

5.3.3 Dihydropyran-4-Ones

The oxa-Diels–Alder reaction of benzaldehyde **115a** with 1-methoxy-3-trimethylsilyloxy-1,3-butadiene (Danishefsky's diene) **121** in the presence of

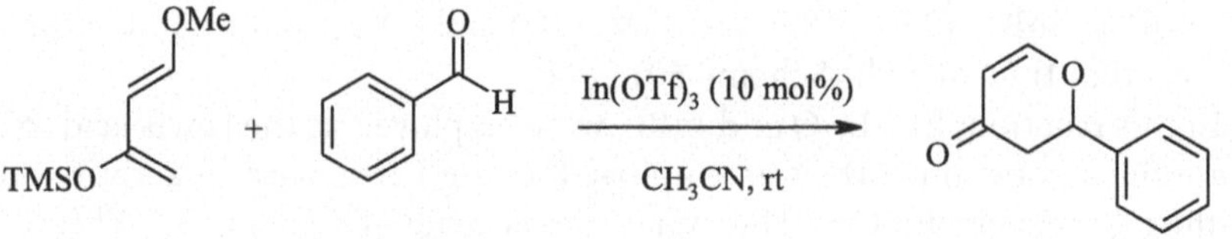

121 **115a** **158**

Scheme 61 Synthesis of pyran-4-ones

10 mol % $In(OTf)_3$ resulted in corresponding pyran-4-ones **158** in 78% within a short time (30 min) at room temperature (Scheme 61) [176].

5.3.4 Pyranopyrimidines and Furopyranopyrimidines

Pyrimidine derivatives continue to be of great interest due to their wide range of biological activities [195]. Preparation of naturally occurring complex molecules containing a uracil ring poses significant synthetic challenges [196]. The development of clinically useful anticancer (5-fluorouracil) [197, 198] and antiviral drugs (AZT, DDC, DDI, BVDU) [199–201] has renewed interest in the synthetic manipulation of uracils [202]. The furo[2,3-*d*]pyrimidine derivatives act as useful sedatives, antihistamines, diuretics, muscle relaxants and antiulcer agents. Furthermore, pyrano [2,3-*d*]pyrimidines also represent broad classes of annelated uracils. A number of compounds having these systems were synthesized and were found to have diverse pharmacological activity [203, 204].

Diastereomeric novel pyrano [2,3-*d*]pyrimidines **161a–e** (*trans*) and **162a–e** (*cis*) have been synthesized via a one-pot three-component reaction of 1,3-dimethyl barbituric acid **159** with an aromatic aldehyde **115a–e** and ethyl vinyl ether **160** in presence of 1 mol % of indium(III) chloride in acetonitrile:water (3:1) at room temperature in excellent overall yields (90–99%). $Sc(OTf)_3$ was also effective in terms of yields but with poor stereoselectivities. This reaction involves an inverse electron-demand Diels–Alder reaction between α,β-ethylenic ketones (formed in situ by the Knoevenagel condensation of 1,3-dimethyl barbituric acid with an aromatic aldehydes), which act as oxygen-heterodienes with ethyl vinyl ether. In all cases the *trans* isomers **161** were the predominant products. This reaction works well for various substitution patterns on the aromatic ring of aldehydes. However, the reaction did not proceed with aliphatic or heterocyclic aldehydes, and the results are summarized in Table 34 (Scheme 62) [205].

The structure of these diastereomers was assigned on the basis of spectroscopic and elemental analysis. The NMR spectra of diastereomers **161a** (*trans*) and **162a** (*cis*) exhibited a resonance as a doublet of doublets attributed to the anomeric proton at $\delta = 5.46$ ($J = 3.0$, and 6.3 Hz) for the

Table 34 $InCl_3$ catalyzed synthesis of pyrano [2,3-*d*]pyrimidines

Entry	Aldehydes (**115**) Ar	Time (h)	Product ratio *trans:cis* (**161:162**)	Overall yield (%) (**161+162**)
a	4-NO_2-C_6H_4-	2.5	75:25	99
b	C_6H_5	3.5	70:30	95
c	4-Cl-C_6H_4-	2.5	70:30	95
d	4-Me-C_6H_4-	3.5	70:30	90
e	4-OMe-C_6H_4-	3.5	70:30	90

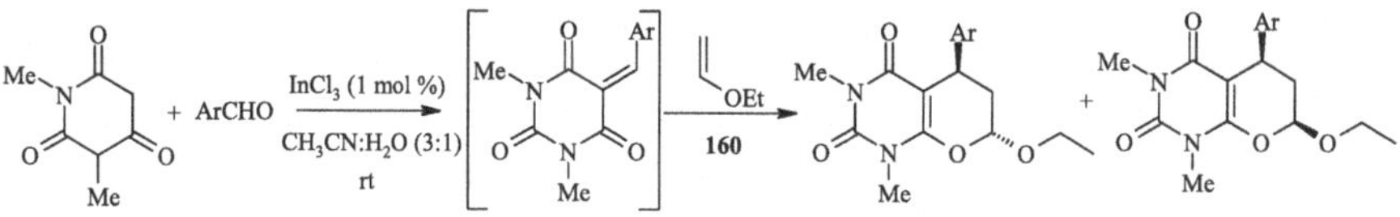

Scheme 62 Synthesis of pyrano [2,3-*d*]pyrimidines

diastereomer **161a** and at $\delta = 5.25$ ($J = 2.1$ and 4.5 Hz) for the diastereomer **162a**, suggesting that the ethoxy substituent occupies a pseudoaxial position within the dihydropyran ring for the diastereomer **161a**. The benzylic proton appeared as a triplet at $\delta = 4.15$ and 4.19 for the diastereomers **161a** and **162a** with an apparent coupling constant $J = 7.5$ and 6.6 Hz, respectively. Analysis of the resonances of the equatorial and axial hydrogens located on the neighboring methylene group of diastereomers **161a** or **162a** indicated, after spin decoupling, that the coupling constants are $J = 7.0$ and 6.1 Hz, respectively, for diastereomer **161a**. Thus, the configuration of diastereomer **161a** may therefore be assigned as *trans* and *cis* for the diastereomer **162a**.

Similarly, diastereomeric novel furopyrano [2,3-*d*]pyrimidines **163a–e** (*cis*) was prepared via a one-pot three-component reaction of 1,3-dimethyl barbituric acid **159** with an aromatic aldehydes **115** and 2,3-dihydrofuran **66** in presence of 1 mol % of indium(III) chloride in acetonitrile:water (3:1) at room temperature in good yields (80–90%). In all cases the reactions

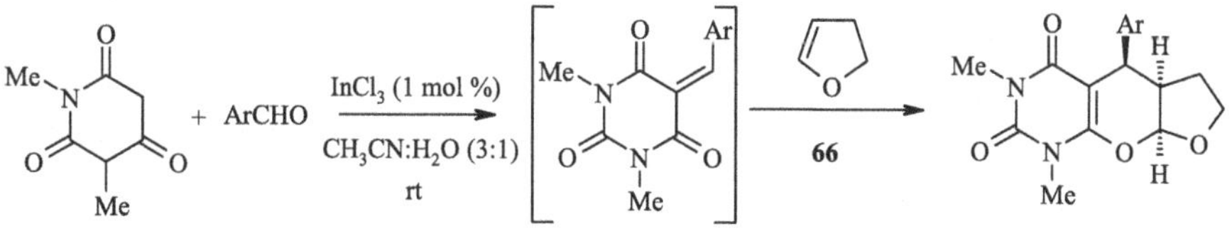

Scheme 63 Synthesis of furopyrano [2,3-*d*]pyrimidines

Table 35 $InCl_3$ catalyzed synthesis of furopyrano [2,3-*d*]pyrimidines

Aldehydes (**115**) Ar	Time (h)	Product **163** (*cis*)	Yield (%)
4-NO_2–C_6H_4–	8.0	**163a**	80
C_6H_5	10.0	**163b**	85
4-Cl–C_6H_4–	8.0	**163c**	90
4-Me–C_6H_4–	8.0	**163d**	80
4-OMe–C_6H_4–	10.0	**163e**	82

proceeded smoothly at ambient temperature with high *cis* selectivity. The tetrahydrofuran ring is *cis*-fused as indicated by the coupling constant of the protons at δ5.75 is $J = 6$ Hz and δ4.48 is $J = 5.4$ Hz. Several examples illustrating this novel and rapid procedure for the synthesis of fused pyrimidines are summarized in Table 35 (Scheme 63) [205].

5.3.5 Pyranobenzopyrans and Furanobenzopyrans

2*H*-1-Benzopyrans (chromenes) and 3,4-dihydro-2*H*-1-benzopyrans (chromans) are important classes of oxygenated heterocycles that have attracted much synthetic interest because of the biological activity of naturally occurring representatives [206, 207]. Numerous 4-amino-benzopyrans and their derivatives have drawn considerable attention in the last decade as modulators of potassium channels influencing the activity of the heart and blood pressure [208, 209]. Particularly, fused tetrahydropyranobenzopyran derivatives are frequently found in naturally occurring bioactive molecules.

Novel *cis*-fused amino-substituted pyranobenzopyrans **164a–d** (*cis*) and **165a–d** (*trans*) have been prepared via a one-pot three-component reaction of *o*-hydroxybenzaldehyde **115b** with an aromatic amine **62** and 3,4-dihydro-2*H*-pyran **55** in the presence of 20 mol % of indium(III) chloride in acetonitrile at room temperature in good overall yields (65–91%); results are presented in Table 36 (Scheme 64) [210].

Triphenylphosphonium perchlorate is also an effective catalyst in terms of yields and selectivities. The ratio of **164** and **165** was determined from the crude ^{1}H NMR spectra of the products. The stereochemistry of product **164** was assigned based on the coupling constants and nOe studies. The six-membered tetrahydropyran ring was *cis*-fused, as indicated by the coupling constant $J_{4-5} = 2.5$ Hz between H_5 (δ5.47) and H_4 for **164b**. Also $J_{4-6} = 1.8$ Hz (H_6, δ4.88) for product **164b** and the presence of an nOe between H_6–H_5 and H_5–H_4, support that H_6 is *cis* to H_4. The product **165b** differs from **164b** having a different configuration at C_6. This was supported by the coupling constants, as well as the absence of an nOe between H_5–H_6 and H_6–H_4.

Table 36 $InCl_3$ catalyzed synthesis of pyranobenzopyrans

Entry	Anilines (**62**) R^1	Time (h)	Product ratio **164**:**165**	Overall yield (%) (**164**+**165**)
a	C_6H_5	1.0	93:07	91
b	4-Me–C_6H_5	1.5	90:10	85
c	4-Br–C_6H_5	1.5	75:25	84
d	4-NO_2–C_6H_5	3.0	70:30	65

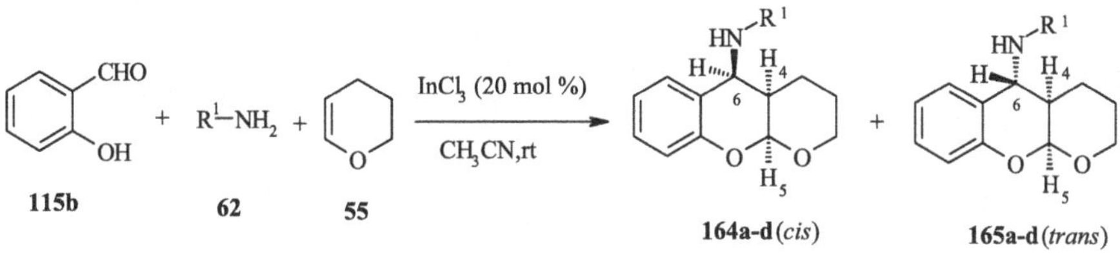

Scheme 64 Synthesis of pyranobenzopyrans

This reaction probably proceeds through the activation of the imine (in situ formed by the reaction of *o*-hydroxybenzaldehyde and aromatic amine) aided by the catalyst followed by addition and subsequent cyclization of the enol ether to afford corresponding benzopyrans (Fig. 20). Alternatively, we also assume that the activation of imine by the catalyst can lead to the formation of *o*-quininomethide **A**, which subsequently undergo [4+2] oxa-Diels–Alder reaction with dienophile enol ether to afford the corresponding benzopyrans (Fig. 21).

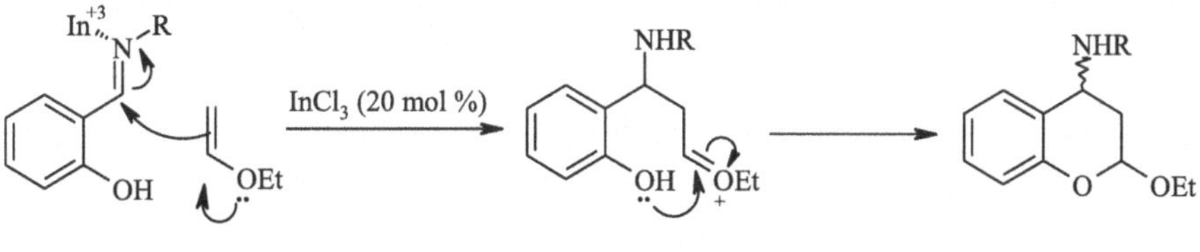

Fig. 20 Step-wise mechanism for the synthesis of benzopyrans

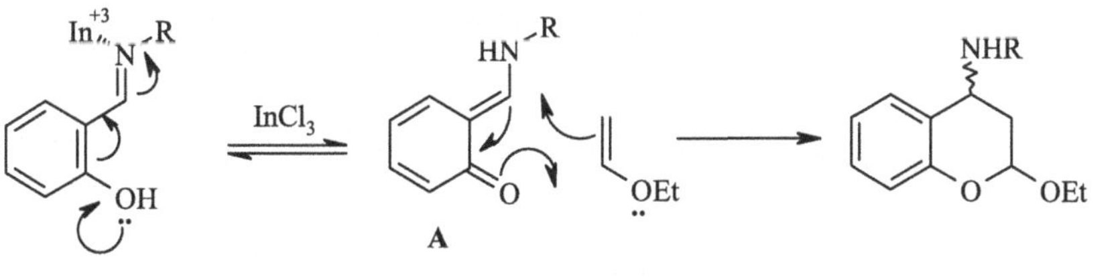

Fig. 21 Concerted mechanism for the synthesis of benzopyrans

Similarly, furanobenzopyrans **166a–c** (*cis*) and **167a–c** (*trans*) have been synthesized via a one-pot three-component reaction of *o*-hydroxybenzaldehyde **115b** with aromatic amines **62** and 3,4-dihydro-2*H*-pyran **66** in the presence of 20 mol % of indium(III) chloride in acetonitrile at room temperature in good overall yields (87–93%) favoring *cis* isomers; results are summarized in Table 37 (Scheme 65) [210].

Table 37 $InCl_3$ catalyzed synthesis of furanobenzopyrans

Entry	Anilines (**62**) R^1	Time (h)	Product ratio **166**:**167**	Overall yield (%) (**166**+**167**)
a	C_6H_5	0.5	85:15	93
b	4-Me–C_6H_5	1.0	80:20	89
c	4-Br–C_6H_5	1.5	78:22	87

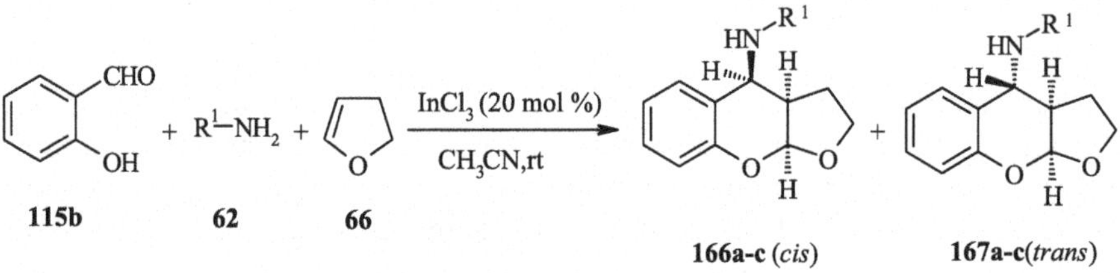

Scheme 65 Synthesis of furanobenzopyrans

Furthermore, the reaction between *o*-hydroxybenzaldehyde **115**, aniline **62a** and ethyl vinyl ether **160** in dichloromethane using 20 mol % $InCl_3$ afforded the corresponding diastereomeric 2-ethoxy-4-*N*-phenylaminobenzopyrans **168** (*cis*) and **169** (*trans*) in high yields and with high *cis* selectivity. However, this reaction did not proceed with amines such as benzyl, cyclohexyl, methyl and ethyl amines used for cyclization (Scheme 66).

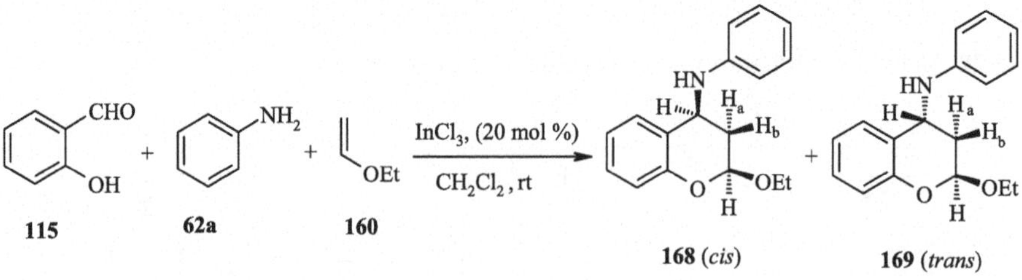

Scheme 66 Synthesis of benzopyrans

5.4
[3+3] Cycloadditions

5.4.1
Bicyclic Tetrahydroquinolines

Bicyclic tetrahydroquinolines have attracted great attention owing to their important biological activities including potent antiviral, antitumor activities [211, 212]. Aminoglycosidation of D-glucal **170** with different electron-rich as well as electron-deficient aryl amines **62** using 10 mol % $InBr_3$ or 1.0 eq TMSOTf in CH_2Cl_2 at room temperature afforded the corresponding bicyclic tetrahydroquinolines **171a–g** in good yields with high stereoselectivity (Scheme 67).

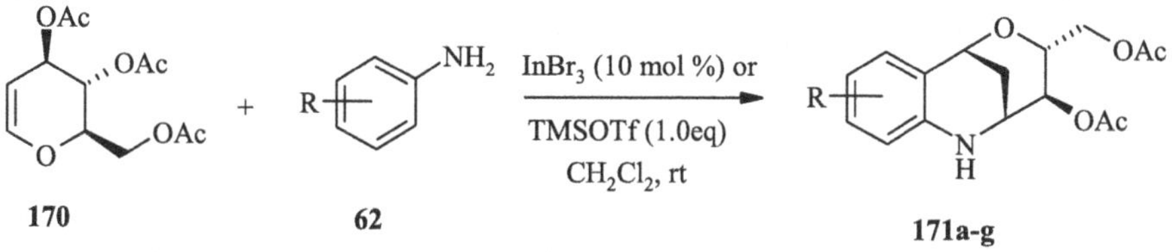

Scheme 67 Synthesis of bicyclic tetrahydroquinolines

Similarly L-rhamnal and D-xylal also underwent cyclization to afford the corresponding bicyclic adducts **171h–k** (see Table 38). Among various Lewis acids, $InBr_3$, TMSOTf, $CeCl_3 \cdot 7H_2O$, YCl_3, $YbCl_3$, $Sc(OTf)_3$, $Bi(OTf)_3$, $Yb(OTf)_3$, $Ce(OTf)_3$, $Sm(OTf)_3$ were screened for this conversion, and among

Table 38 $InBr_3$ or TMSOTf catalyzed synthesis of bicyclic tetrahydroquinolines

Glucal	Amines (**62**) R	Product	$InBr_3$ (10 mol %) Time (h)	Yield (%)	TMSOTf (1.0 eq) Time (h)	Yield (%)
D-glucal	H	**171a**	6.0	85	3.5	87
D-glucal	2-Me	**171b**	5.5	82	4.0	85
D-glucal	4-F	**171c**	7.0	78	5.0	81
D-glucal	4-Br	**171d**	7.5	82	4.5	84
D-glucal	4-OMe	**171e**	7.0	78	4.0	78
D-glucal	4-Cl	**171f**	6.0	84	4.5	85
D-glucal	2,6-$(CH_3)_2$	**171g**	8.0	–	6.0	–
L-rhamnal	H	**171h**	5.0	87	5.0	85
L-rhamnal	4-Me	**171i**	4.5	82	4.0	84
D-xylal	H	**171j**	6.0	89	3.5	82
D-xylal	4-Me	**171k**	5.0	85	4.0	80

them $InBr_3$ and TMSOTf were found to be superior in terms of yields and stereoselectivites [213].

The product **171a** was characterized thoroughly by various NMR experiments including double quantum filtered correlation spectroscopy (DQF-COSY), nuclear Overhauser effect spectroscopy (NOESY), heteronuclear single quantum correlation spectroscopy (HSQC), and $3J_{CH}$-optimized HMBC experiments. The edited HSQC spectrum showed the presence of two methylene groups in addition to eight methine and two methyl groups. The location of the methylene group in the bridge of a bicyclononene-like structure was confirmed by the presence of small couplings between these protons and the bridgehead protons H_1 and H_3 ($J_{H1\text{-}H2(pro\text{-}S)} = 3.7$ Hz, $J_{H1\text{-}H2(pro\text{-}R)} = 1.8$ Hz, $J_{H2(pro\text{-}S)\text{-}H3} = 2.4$ Hz, and $J_{H2(pro\text{-}R)\text{-}H3} = 4.6$ Hz) (Fig. 22). Fusion of the bicyclononene and the aromatic ring at C_1–C_{11} and NH–C_3 was confirmed by nOe interactions between H_1 and H_{12}. Further support for the structure came from HMBC peaks for H_1/C_{12}, H_1/C_{11}, H_1/C_{16} and H_{12}/C_1. The two six-membered rings of the bicyclononane moiety have two different conformations. The one containing oxygen takes a chair form, whereas the unsaturated ring with nitrogen and fused to the aromatic ring exists in a half-chair form. HMBC peaks for $H_{2(pro\text{-}S)}/C_{11}$ and $H_{2(pro\text{-}R)}/C_4$ were consistent with this structure. The large coupling constant $J_{H4\text{-}H5} = 10.4$ Hz (diaxial) and the NOESY cross peak for $H_{2(pro\text{-}S)}$/H4 further support the chair form for the ring containing these protons. The ring current of the aromatic ring causes high-field chemical shifts for $H_{2(pro\text{-}R)}$ ($\delta = 1.96$ ppm) and H5 ($\delta = 3.58$ ppm). Furthermore, the structure of **171a** was confirmed by molecular mechanics calculations [214, 215].

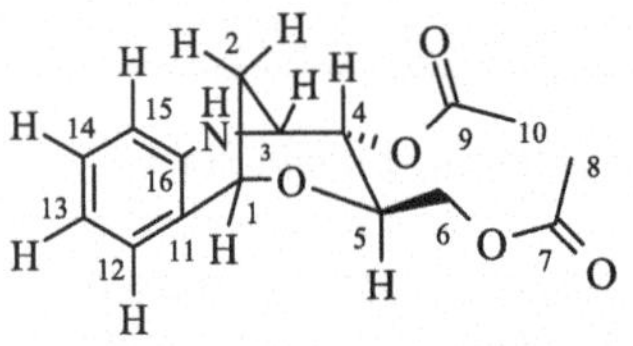

a

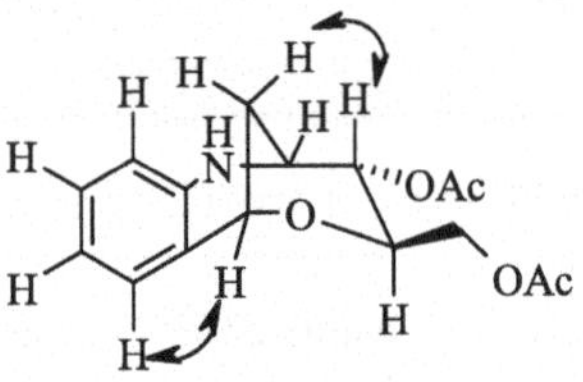

b

Fig. 22 **a** Chemical structure of **171a**. **b** Characteristic nOe interactions for **171a**

A possible mechanism for the formation of bicyclic tetrahydroquinolines **171a–k** is depicted in Fig. 23. To elucidate the mechanistic pathway, the reaction with deuterated aniline **62a′** and 3,4,6-tri-*O*-acetyl-D-glucal **170** was carried out and there was no deuterium incorporation observed in the product. However, when the reaction of aniline **62a** and **170** in D_2O at 80 °C gave the isomeric deuterated products [2pro-R-D1]-**172a** and [2pro-S-D1]-**172a** in equal amounts, which was confirmed by ^1H NMR and FAB mass spectroscopy, this clearly indicated that protons were abstracted from the solvent

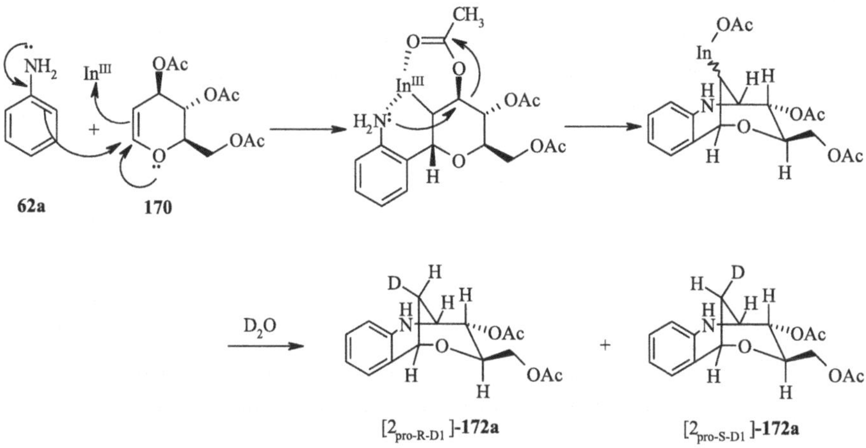

Fig. 23 Possible mechanism for the formation of bicyclic tetrahydroquinolines

and not from aniline. It was also found that **170** does not react with 2,6-dimethylaniline **62** (R=2,6-$(CH_3)_2$) (Table 38), which indicated that one of the *ortho*-positions of aniline should be free from substitution for the success of the reaction.

In a similar fashion, bicyclic tetrahydroquinolines **171a–e** have been synthesized by the reaction of anilines **62** with optically active 4,6-di-*O*-acetyl-2,3-dideoxyaldehydo-D-erythro-*trans*-hex-2-enose **173** in the presence of 10 mol % of $InCl_3$ in CH_3CN at room temperature in excellent yields with high stereoselectivity (Scheme 68) [216].

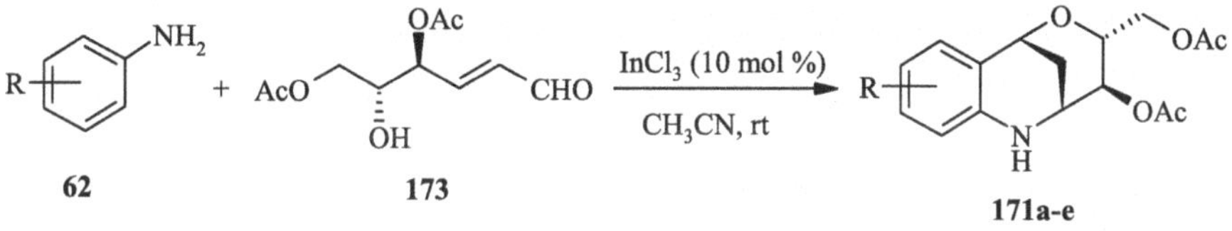

Scheme 68 Synthesis of bicyclic tetrahydroquinolines

However, *ortho*-hydroxy-substituted anilines and *trans*-cinnamaldehyde did not afford any bicyclic adducts. This reaction was successful only with δ-hydroxy-α,β-unsaturated aldehydes. This reaction was also carried out effectively with 10 mol % of $Yb(OTf)_3$ under mild conditions.

The probable mechanism was depicted in Fig. 24. Initially Michael addition of aniline **62a** to the unsaturated position of conjugated aldehydes **173** afforded an adduct, which may form an *o*-quinoid intermediate by loss of water followed by intramolecular attack by an OH function leading to the formation of bicyclic adduct **171a**.

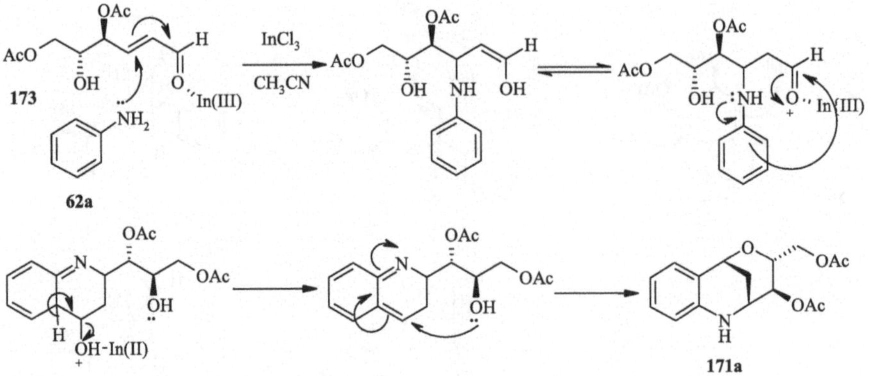

Fig. 24 Possible mechanism for the synthesis of bicyclic tetrahydroquinolines

References

1. Bertz SH (1981) J Am Chem Soc 103:3599
2. Huisgen R (1968) Angew Chem Int Ed Engl 7:321
3. Woodward RB, Hoffmann R (1969) Angew Chem Int Ed Engl 8:781
4. Woodward RB, Hoffmann R (1969) Angew Chem 81:797
5. Frost CG, Chauhan KK (2000) J Chem Soc Perkin Trans 1:3015
6. Kobayashi S, Manabe K (2000) Pure Appl Chem 72:1373
7. Bandini M, Cozzi G, Melchiorre P, Umani-Ronchi A (2001) Tetrahedron Lett 42:3041
8. Kobayashi S, Manabe K (2002) Acc Chem Res 35:209
9. Zhang J, Li C-J (2002) J Org Chem 67:3969
10. Yadav JS, Reddy BVS, Raju AK, Rao CV (2002) Tetrahedron Lett 43:5437
11. Yadav JS, Reddy BVS, Satheesh G, Prabhakar A, Kunwar AC (2003) Tetrahedron Lett 44:2221
12. Yadav JS, Reddy BVS, Parimala MS (2003) Synthesis 2390
13. Yadav JS, Reddy BVS, Swamy T (2004) Synthesis 106
14. Yadav JS, Reddy BVS, Aravind S, Kumar GGKSN, Reddy AS (2007) Tetrahedron Lett 48:6117
15. Dorbec M, Florent J-C, Monneret C, Rager MN, Bertounesque E (2007) Tetrahedron 62:11766
16. Shanthi G, Subbulakshmi G, Perumal PT (2007) Tetrahedron 63:2057
17. Ghosh R, Chakraborthy A, Maiti DK, Puranik VG (2006) Org Lett 8:1061
18. Dey S, Nandi D, Pradhan PK, Giri VS, Jaisankar P (2007) Tetrahedron Lett 48:2573
19. Lee C-HA, Loh T-P (2006) Tetrahedron Lett 47:1641
20. Sakai N, Annaka K, Konakahara T (2006) Tetrahedron Lett 47:631
21. Sakai N, Annaka K, Konakahara T (2006) J Org Chem 71:3653
22. Trivedi R, De SK, Gibbs RA (2006) J Mol Catal A Chem 245:8
23. Ghosh R, Maiti S, Chakraborthy A (2004) J Mol Catal A Chem 217:47
24. Lautens M, Klute W, Tam W (1996) Chem Rev 96:49
25. Hauf H-W (1997) Chem Rev 97:523
26. Gothelf KV, Jorgensen KA (1998) Chem Rev 98:863
27. Coldham I, Hufton R (2005) Chem Rev 105:2765

28. Pandey G, Banerjee P, Gadre SR (2006) Chem Rev 106:4484
29. Pearson WH, Lian BW, Bergmeier SC (1996) In: Padwa A (ed) Comprehensive Heterocyclic Chemistry II, vol 1A. Pergamon, New York, p 1
30. Tanner D (1994) Angew Chem Int Ed Engl 33:599
31. McCoull W, Davis FA (2000) Synthesis 1347
32. Sengupta S, Mondal S (2000) Tetrahedron Lett 41:6245
33. Bartnik R, Mloston G (1983) Synthesis 924
34. Jephcote VJ, John DI, Williams DJ (1986) J Chem Soc Perkin Trans I:2195
35. Palacios F, Ochoa de Retana AM, Alonso JM (2005) J Org Chem 70:8895
36. Quin LD (2000) A Guide to Organophosphorus Chemistry. Wiley Interscience, New York
37. Kukhar VP, Hudson HR (eds) (2000) Aminophosphonic and Aminophosphinic Acids. Chemistry and Biological Activity. Wiley, Chichester, UK
38. Mader MM, Bartlett PA (1997) Chem Rev 97:1281
39. Pellicciari R, Amori L, Kuznetsova N, Zlotsky S, Gioiello A (2007) Tetrahedron Lett 48:4911
40. Drescher M, Li Y-F, Hammerschmidt F (1995) Tetrahedron 51:4933
41. Drag M, Grembecka J, Pawelczak M, Kafarski P (2005) Eur J Med Chem 40:764
42. Iron A, Ruart M, Duboy JP, Beranger M, Cassaigne A, Neuzil E (1981) Biochem Soc Trans 9:246
43. Iron A, Covi G, Beranger M, Cassaigne A, Neuzil E (1986) Biochem Soc Trans 14:641
44. Dembitsky VM (2008) J Nat Med 62:1
45. Yadav JS, Reddy BVS, Kondaji G (2003) Synthesis 7:1100
46. Jensen KB, Hazell RG, Jorgensen KA (1999) J Org Chem 64:2353
47. Gothelf KV, Jorgensen KA (1998) Chem Rev 98:863
48. Xie Z-F, Mo X-X, Mo X-L, Liu F-M (2007) Acta Cryst E63:o3010
49. Curtius T (1883) Ber Dtsch Chem Ges 16:2230
50. Buchner E (1888) Ber Dtsch Chem Ges 21:2637
51. Huisgen R (1984) In: Padwa A (ed) 1,3-Dipolar-Cycloaddition Chemistry, vol 1. Wiley, New York, p 1
52. Padwa A (1991) In: Trost BM, Fleming I (eds) Comprehensive Organic Synthesis, vol 4. Pergamon, Oxford, p 1069
53. Wade PA (1991) In: Trost BM, Fleming I (eds) Comprehensive Organic Synthesis, vol 4. Pergamon, Oxford, p 1111
54. Grigg R, Jones MF, Sridharan V (2001) Tetrahedron 57:7979
55. Apers S, Vlietinck A, Pieters L (2003) Phytochem Rev 2:201
56. Ward RS (1999) Nat Prod Rep 16:75
57. Engler TA, Wei D, Letavic MA, Combrink KD, Reddy PJ (1994) J Org Chem 59:6588
58. Shen TY, Hwang SB, Chang MN, Doebber TW, Lam MH, Wu MS, Wang X (1985) Int J Tissue React 7:339
59. Pieters L, Van Dyck S, Gao M, Bai R, Hamel E, Vlietinck A, Lemiere G (1999) J Med Chem 42:5475–5481
60. Pessini GL, Filho BPD, Nakamura CV, Cortez DAG (2005) J Braz Chem Soc 16:1130
61. Chakravarty AK, Mukhopadhyay S, Saha S, Pakarashi SC (1996) Phytochemistry 41:935
62. Cho JY, Baik KU, Yoo ES, Yoshikawa K, Park MH (2000) J Org Chem 63:1205
63. Micard V, Grabber JH, Ralph J, Renardt CMGC, Thibault J-F (1997) Phytochemistry 44:1365
64. Parlow JJ (1998) J Heterocycl Chem 35:1493
65. Finkelstein BL, Strock CJ (1997) Pestic Sci 50:324

66. Daidone G, Maggio B, Plescia S, Raffia D, Musiu C, Milia C, Perra G, Marongiu ME (1998) Eur J Med Chem 33:375
67. Nauduri D, Reddy GB (1998) Chem Pharm Bull 46:1254
68. Jiang N, Li C-J (2004) Chem Commun 394
69. Fleming I (1976) Frontier Orbitals and Organic Chemical Reactions. Wiley, Chichester
70. Bertozzi CR, Kiessling LL (2001) Science 291:2355
71. Rudd PM, Elliott T, Cresswell P, Wilson IA, Dwek RA (2001) Science 291:2370
72. Pandey N, Vasella A (2000) Helv Chim Acta 83:1205
73. Mitchell ML, Tian F, Lee LV, Wong C-H (2002) Angew Chem Int Ed 41:3041
74. Lenda F, Guenoun F, Martineza J, Lamatya F (2007) Tetrahedron Lett 48:805
75. Yanai H, Taguchi T (2005) Tetrahedron Lett 46:8639
76. Dewick PM (1997) Medicinal Natural Products, chap 6. Wiley, Chichester, UK
77. Dean FM, Sargent MV (1984) In: Katrizky AR, Rees CW (eds) Comprehensive Heterocyclic Chemistry, vol 4. Pergamon, Oxford, UK, p 531
78. Morikawa S, Yamazaki S, Furusaki Y, Amano N, Zenke K, Kakiuchi K (2006) J Org Chem 71:3540
79. Yamazaki S, Ohmitsu K, Ohi K, Otsubo T, Moriyama K (2005) Org Lett 7:759
80. Diels O, Alder K (1928) Justus Liebigs Ann Chem 460:98
81. Woodward RB, Katz TJ (1959) Tetrahedron 5:70
82. Woodward RB, Hoffmann R (1965) J Am Chem Soc 87:395
83. Woodward RB, Hoffmann R (1970) The Conservation of Orbital Symmetry. Verlag Chemie, Weinheim
84. Prajapathi D, Laskar DD, Sandhu JS (2000) Tetrahedron Lett 41:8639
85. Yanai H, Saito A, Taguchi T (2005) Tetrahedron 61:7087
86. Daly JW, Spande TF (1986) In: Pelletier SW (ed) Alkaloids: Chemical and Biological Perspectives, vol 4. Wiley, New York, p 1
87. Fodor GB, Colasanti B (1985) In: Pelletier SW (ed) Alkaloids: Chemical and Biological Perspectives, vol 3. Wiley, New York, p 1
88. Bogor DL, Weinreb SM (1987) Hetero Diels–Alder Methodology in Organic Synthesis, chap 2. Academic, San Diego, p 34
89. Bogor DL, Weinreb SM (1987) Hetero Diels–Alder Methodology in Organic Synthesis, chap 9. Academic, San Diego, p 239
90. Bogor DL, Patel M (1989) In: Suschitzky H, Scriven EFV (eds) Progress in Heterocyclic Chemistry. Pergamon, New York, p 1
91. Kametani T, Hibino S (1987) Adv Heterocycl Chem 42:246
92. Povarov LS (1967) Russian Chem Rev 36:656
93. Barluenga J, Tomas M (1993) Adv Heterocycl Chem 57:1
94. Xu J-Y (1995) Youji Huaxue 15:133
95. Boger DL (1986) Chem Rev 86:781
96. Boger DL (1983) Tetrahedron 39:2869
97. Buonora P, Olsen J-C, Oh T (2001) Tetrahedron 57:6099
98. Babu G, Perumal PT (1997) Tetrahedron Lett 38:5025
99. Babu G, Nagarajan R, Natarajan R, Perumal PT (2000) Synthesis 5:661
100. Nandagopal S, Geeta B, Perumal PT (2004) Syn Commun 34:1157
101. Anzini M, Cappelli A, Vomero S, Cagnotto A, Skorupska M (1993) Med Chem Res 3:44
102. Quraishi MA, Thakum VR, Dhawan SN (1989) Indian J Chem Sec B 28B:891
103. Qkazaki S, Asao T, Wakida M, Jshida K, Washinosu M, Utsugi T, Yamada Y (1996) Chem Abstr 124:232268a

104. Brooks JR, Berman C, Hichens M, Primka RL, Reynolds GF, Ramusson GH (1982) Proc Soc Exp Biol Med 169:67
105. Babu G, Perumal PT (1998) Tetrahedron 54:1627
106. Babu G, Perumal PT (1998) Tetrahedron 39:3225
107. Gribble GW, Scriven EFV, Bird CW (1996) In: Katritzky AR, Ress CW (eds) Comprehensive Heterocyclic Chemistry II, vol 2. Pergamon, Oxford, p 207
108. Rykowski A, Lipinska T (1996) Synth Commun 26:4409
109. Eguchi S (2005) ARKIVOC ii:98
110. Nagarajan R, Perumal PT (2002) Tetrahedron 58:1229
111. Sankaranarayanan R, Velmurugan D, Shanmugasundararaj S, Fun HK, Babu G, Perumal PT (2000) Acta Cryst C 56:475
112. Dieterle F, Muller-Hagedon S, Liebich HM, Gauglitz G (2003) Artif Intell Med 28:265
113. Rowley DC, Hansen MST, Rhodes D, Striffer CA, Ni H, McCammon JA, Bushman FD, Fenical W (2002) Bioorg Med Chem 10:3619
114. Kumar A, Sinha S, Chauhan PMS (2002) Chem Lett 12:667
115. Bher J-B, Gourlain T, Helimi A, Guillerm G (2003) Bioorg Med Chem Lett 13:1713
116. Hosni HM, Basyouni WM, El-Nahas HA (1999) J Chem Res (S):646
117. Botta M, Occhionero F, Nicoletti R, Mastromarino P, Conti C, Margini M, Saladino J (1999) Bioorg Med Chem 9:1925
118. Bhargava G, Kumar V, Mahajan MP (2007) Tetrahedron Lett 48:2365
119. Babu G, Perumal PT (1999) Tetrahedron 55:4793
120. Johnson JV, Rauckman S, Baccanari PD, Roth BJ (1989) J Med Chem 32:1942
121. Yamada N, Kadowaki S, Takahashi K, Umezu K (1992) Biochem Pharmacol 44:1211
122. McLaughlin MJ, Hsung RP (2001) J Org Chem 66:1049
123. Michael JP (2005) Nat Prod Rep 22:627
124. Yadav JS, Reddy BVS, Rao RS, Kumar SK, Kunwar AC (2002) Tetrahedron 58:7891
125. Hudson JB, Graham EA, Chan GC, Towers GHN (1985) Photochem Photobiol 42:523
126. Mizuta M, Kannamori H (1985) Mutat Res 144:221
127. Aguinaldo AM, Dalangin-Mallari VM, Macabeo APG (2007) Int J Antimicrob Agents 29:738
128. McCommick JL, McTee TC, Cardellina JH, Boyd MR (1996) J Nat Prod 59:469
129. Zhang J, Li C-J (2002) J Org Chem 67:3969
130. Woods GF (1955) Organic Syntheses, vol III. Wiley, New York, p 470
131. Schniepp LE, Geller HH (1946) J Am Chem Soc 68:1646
132. Kantam ML, Roy M, Roy S, Subhas MS, Sreedhar B, Choudary BM, De RL (2007) J Mol Catal A Chem 265:244
133. Badawey ESAM, Kappe T (1997) Eur J Med Chem 32:815
134. Leach CA, Brown TH, Ife RJ, Keeling DJ, Laing SM, Parsons ME, Price CA, Wiggall KJ (1992) J Med Chem 35:1845
135. Brown TH, Ife RJ, Keeling DJ, Laing SM, Leach CA, Parsons ME, Price CA, Revalli DR, Wiggall KJ (1990) J Med Chem 33:527
136. Witherup KM, Ransom RW, Graham AC, Bernard AM, Salvatore MJ, Lumma WC, Anderson PS, Pitzenberger SM, Varga SL (1995) J Am Chem Soc 117:6682
137. Hadden M, Stevenson PJ (1999) Tetrahedron Lett 40:1215
138. Hadden M, Nieuwenhuyzen M, Potts D, Stevenson PJ, Thompson N (2001) Tetrahedron 57:5615
139. Williams DH, Fleming I (1995) Spectroscopic Methods in Organic Chemistry, vol 5. McGraw Hill, London
140. Elamparuthi E, Anniyappan M, Muralidharan D, Perumal PT (2005) ARKIVOC xi:6
141. Gadhwal S, Sandhu JS (2000) J Chem Soc Perkin Trans I:2827

142. Skotnicki JS, Steinbaugh BA, Fitzgerald JJ, Kearney RM, Mosser JH, Adams LM, Caccere RG, Chang JY, Gilman SC (1991) Med Chem Res 1:245
143. Graul AI (1997) Drugs News Perspect 10:24
144. Bertenshaw SR, Talley JJ, Rogier DJ, Graneto MJ, Koboldt CM, Zhang Y (1996) Bioorg Med Chem Lett 6:2827
145. Rathna RSMD, Jayashankaran J, Ramesh R, Raghunathan R (2006) Tetrahedron Lett 47:7571
146. Rathna RSMD, Jayashankaran J, Raghunathan R (2007) Tetrahedron Lett 48:4139
147. Sridharan V, Perumal PT, Avendano C, Menendez JC (2007) Org Biomol Chem 5:1351
148. Jimenez O, Rosa GDL, Lavilla R (2005) Angew Chem Int Ed 44:6521
149. Christoffers J (2005) In: Baro A (ed) Quaternary Stereocenters: Challenges and Solutions for Organic Synthesis. Wiley-VCH, Weinheim
150. Duvelleroy D, Perrio C, Parisel O, Lasne M-C (2005) Org Biomol Chem 3:3794
151. Sliwa W (1978) Studies on Benzo[*h*]naphthyridines, Scientific Papers of the Institute of Organic and Physical Chemistry. No. 13/8, Wroclaw Technical University, Wroclaw
152. Matusiak G, Sliwa W (1988) Acta Chim Hung 125:267
153. Chrząstek L, Mianowska B, Sliwa W (1994) Aust J Chem 47:2129
154. Zelichowicz N, Gaudyn A (1992) Chem Papers 46:284
155. Lavilla R, Bernabeu MC, Carranco I, Diaz JL (2003) Org Lett 5:717
156. Huang T, Li CJ (2000) Tetrahedron Lett 41:6715
157. Michael JP (2001) Nat Prod Rep 18:520
158. Barluenga J, Mateos C, Aznar F, Valdes C (2002) Org Lett 4:1971
159. Buschmann N, Ruckert A, Blechert S (2002) J Org Chem 67:4325
160. Vlietinck AJ, Bruyne TD, Apers S, Pieters LA (1998) Planta Med 64:97
161. Dragovich PS, Zhou R, Prins TJ (2002) J Org Chem 67:741
162. Amat M, Bosch J, Hidalgo J, Canto M, Perez M, Llor N, Molins E, Miravitlles C, Orozco M, Luque J (2000) J Org Chem 65:3074
163. Amat M, Llor N, Bosch J, Solans X (1997) Tetrahedron 53:719
164. Alcaide B, Almendros P, Alonso JM, Aly MF (2003) Chem Eur J 9:3415
165. Badorrey R, Cativiela C, Diaz-de-Villegas MD, Galvez JA (1999) Tetrahedron 55:7601
166. Yadav JS, Reddy BVS, Rao CV (2002) Synlett 993
167. Yadav JS, Reddy BVS, Sadasiv K, Reddy PSR (2002) Tetrahedron Lett 43:3853
168. Yadav JS, Reddy BVS, Madhuri CR, Sabitha G (2001) Synthesis 1065
169. Laurent-Robert H, Garrigues B, Dubac J (2000) Synlett 1160
170. Hermitage S, Jay DA, Whiting A (2002) Tetrahedron Lett 43:9633
171. Kobayashi S, Ishitani H, Nagayama S (1995) Synthesis 1195
172. Zhang J, Li C-J (2002) J Org Chem 67:3969
173. Sabitha G, Reddy EV, Yadav JS, Krishna KVSR, Sankar AR (2002) Tetrahedron Lett 43:4029
174. Watson PS, Jiang B, Scott B (2000) Org Lett 2:3679
175. Clarke PA, Zaytzev AV, Whitewood AC (2007) Tetrahedron Lett 48:5209
176. Ali T, Chauhan KK, Frost CG (1999) Tetrahedron Lett 40:5621
177. Birkinshaw TN, Tabor AB, Holmes AB, Kaye P, Mayne PM, Raithby PR (1988) J Chem Soc Chem Commun 1599
178. Birkinshaw TN, Tabor AB, Holmes AB, Raithby PR (1988) J Chem Soc Chem Commun 1601
179. Birkinshaw TN, Tabor AB, Holmes AB, Kaye P, Mayne PM, Raithby PR (1988) J Chem Soc Chem Commun 1599
180. Osborn HMI, Coisson D (2004) Mini Rev Org Chem 1:41

181. Jorgensen KA (2000) Angew Chem Int Ed 39:3558
182. Nicolaou KC, Sorensen EJ (1996) Classics in Total Synthesis: Targets, Strategies, Methods. VCH, Weinheim
183. Tietze LF (1998) Curr Org Chem 2:19
184. Tietze LF, Pfeiffer T, Schuffenhaur A (1998) Eur J Org Chem 2733
185. Robak J, Gryglewski RJ (1996) Pol J Pharmacol 48:55
186. Wang HK, Morris-Natschke SL, Lee KH (1997) Med Res Rev 17:367
187. Rukavishnikov AV, Smith MP, Birrell GB, Keana JFW, Griffith OH (1998) Tetrahedron Lett 39:6637
188. Garcia A, Paz S, Dominguez D (2001) Tetrahedron Lett 42:665
189. Neighbors JD, Salnikova MS, Beutlerc JA, Wiemer DF (2006) Bioorg Med Chem 14:1771
190. Anniyappan M, Muralidharan D, Perumal PT, Vittal JJ (2004) Tetrahedron 60:2965
191. Jorgensen KA (2000) Angew Chem Int Ed 39:3558
192. Faul MM, Huff BE (2000) Chem Rev 100:2407
193. Harris JM, O'Doherthy GA (2000) Tetrahedron Lett 41:183
194. Liu J, Li X, Wang J, Feng X (2006) Adv Synth Catal 348:939
195. Jones AS, Swgers JR, Walker RT, Clercq ED (1988) J Med Chem 31:268
196. Brown JD (1984) In: Katritzky AR, Rees CW (eds) Comprehensive Heterocyclic Chemistry, vol 3. Pergamon, Oxford, p 57
197. Heidelberger C, Arafield FJ (1964) Cancer Res 23:1226
198. Heidelberger C, Arafield FJ (1964) Chem Abstr 60:2197
199. Clercq ED (1986) J Med Chem 29:1561
200. Baba M, Pauvels R, Herdwig P, Clercq ED, Desmyster J, Vandepulfe M (1987) Biochem Biophys Res Commun 142:128
201. Clercq ED (1986) Anticancer Res 6:549
202. Hirota K, Kubo K, Sajiki H, Kitade Y, Sako M, Maki Y (1997) J Org Chem 62:2999
203. Anderson GL, Shim JL, Brown AD (1976) J Org Chem 41:1095
204. Grivaky EM, Lee S, Siyal CW, Duch DS, Nichol CA (1980) J Med Chem 23:327
205. Prajapati D, Gohain M (2006) Beilstein J Org Chem 2:11
206. Schweizer EE, Meeder-Nycz O (1977) In: Ellis GP (ed) Chromenes, Chromanes, Chromones. Wiley-Interscience, New York, p 11
207. Hepworth J (1984) In: Katrizky AR, Rees CW (eds) Comprehensive Heterocyclic Chemistry, vol 3. Pergamon, Oxford, p 737
208. Buckle DR, Arch JRS, Fenwick AE, Hough-Frydrych CSV, Pinto IL, Smith DG, Taylor SG, Tedder JM (1990) J Med Chem 33:3028
209. Rovnyak GC, Ahmed SZ, Ding CZ, Dzwonczyk S, Ferrara FN, Humphreys WG, Grover GJ, Santafianos D, Atwal KS, Baird AJ, McLaughin LG, Normandin DE, Sleph PG, Traeger SC (1997) J Med Chem 40:24
210. Anniyappan M, Muralidharan D, Perumal PT (2002) Tetrahedron 58:10301
211. Mingeot-Leclercq MP, Glupczynski Y, Tulkens PM (1999) Antimicrob Agents Chemother 43:727
212. Dianez MJ, Galan J, Gomez-Sanchez A, Lopez-Castra A, Rico M (1987) J Chem Soc Perkin Trans 1:581
213. Yadav JS, Reddy BVS, Rao KV, Saritha Raj K, Prasad AR, Kiran Kumar S, Kunwar AC, Jayaprakash P, Jagannath B (2003) Angew Chem Int Ed 42:5198
214. Goto H, Osawa EJ (1993) J Chem Soc Perkin Trans 2:187
215. Serena Software (2007) PC Model. Serena Software, Bloomington, IN, USA
216. Yadav JS, Reddy BVS, Padmavani B (2004) Synthesis 405

Top Heterocycl Chem (2008) 13: 173–200
DOI 10.1007/7081_2007_105

Published online: 24 January 2008

[8+2] Cycloaddition Reactions in the Construction of Heterocycles

Vijay Nair[1,2] (✉) · K. Gopalakrishnan Abhilash[1]

[1]Organic Chemistry Section, Regional Research Laboratory (CSIR),
Industrial Estate P.O., 695 019 Trivandrum, India
vijaynair_2001@yahoo.com

[2]Jawaharlal Nehru Center for Advanced Scientific Research, Jakkur P.O.,
560064 Bangalore, India

Abstract [8+2] Cycloaddition reactions leading to the synthesis of heterocyclic systems are summarized. The heteroanalogs of heptafulvene, viz. tropone, tropothione and azaheptafulvenes are the most thoroughly investigated systems in this regard. They provide an easy access to (7,5)-fused heterocyclic systems. A few other systems like indolizines, benzothiete and dienylisobenzofurans have also been shown to deliver novel heterocyclic systems through [8+2] cycloaddition reactions.

Keywords [8+2] Cycloadditions · Azaheptafulvenes · Dienyl isobenzofurans · Heptafulvenyl systems · Indolizines · Tropone · Tropothione

1 Introduction

The cycloaddition reaction between two polyene components is usually referred to as a higher-order cycloaddition reaction. In practice, any cycloaddition reaction that involves more than 6π electrons can be regarded as belonging to the higher-order variant. The importance of this particular synthetic strategy rests mainly on two factors. Firstly, higher-order cycloaddition reactions provide an easy entry into medium-sized ring systems which are difficult to obtain by other means. Secondly, substrates possessing an extended π-array can exhibit different modes of reactivity in cycloaddition reactions depending on the substrate and conditions. Thus, a detailed study of such a system can provide valuable information regarding various aspects related to the stereo- and periselectivity in cycloaddition reactions, in a general sense. Unlike the Diels–Alder reaction, higher-order cycloaddition reactions generally proceed with modest chemical efficiency. This can be attributed to the multiple reactivity profiles that the polyene component can exhibit, thus resulting in a lack of periselectivity.

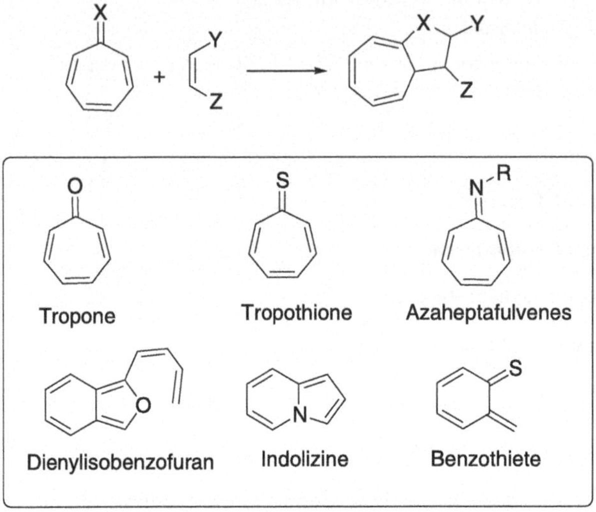

Scheme 1 General representation of [8+2] cycloaddition reaction with the heptafulvenyl system and the structures of commonly encountered 8π systems

[8+2] Cycloaddition reactions represent the most important class belonging to the category of higher-order cycloaddition reactions (Scheme 1). In some cases the products are formally [8+2] adducts but may result from other initially formed cycloaddition intermediates. Heptafulvenes and their heteroanalogs are the most-studied substrates as far as [8+2] cycloaddition reactions are concerned. These reactions provide a direct entry into diversely functionalized bicyclo[5.3.0] ring systems which are part of numerous natural and non-natural compounds. Thus, the hetero-analogs of heptafulvene, viz., tropone, tropothione and azaheptafulvenes, lead to the synthesis of fused (7,5) heterocycles through [8+2] cycloaddition reactions. Indolizines are another class of compounds that possess a hetero-8π system suitable for cycloaddition reactions. Recently, it was reported that dienylfurans and dienylisobenzofurans undergo efficient [8+2] cycloaddition reactions with electron-deficient 2π systems leading to the synthesis of oxabridged macrocycles, a framework found in a number of biologically active molecules. The cycloaddition reactions of benzothiete and benzobisthiete have been utilized for the synthesis of three-dimensional molecular frameworks including bent-shaped polymers. The cycloaddition reactions of each category will be discussed in detail in the following sections.

2
Reactions of Tropone

Among the heptafulvenoids, the cycloaddition chemistry of tropone has been explored in great detail. The studies reveal the preference for tropone to react as a 4π or 6π component in cycloaddition reactions. Nevertheless, there are a few cycloaddition reactions in which tropone takes part as an 8π component. Thus, reaction of tropone with dichloroketene generated in situ from dichloroacetyl chloride and triethylamine afforded the adduct **4**, formally an [8+2] adduct (Scheme 2) [1]. Similarly, diphenylketene produces adduct **3**, in this case via a [2+2] adduct **2**. In contrast, in the reaction of dichloroketene with tropone, formation of a [2+2] adduct was not observed even at low temperatures. Thus, it was proposed that an unstable [2+2] adduct may generate a second ketene **5** or that the reaction may proceed through a nucleophilic attack of the tropone oxygen on the ketene carbonyl (see **6** and **7**) [1]. It is interesting to note that the potential [8+2] cycloaddition pathway was not considered.

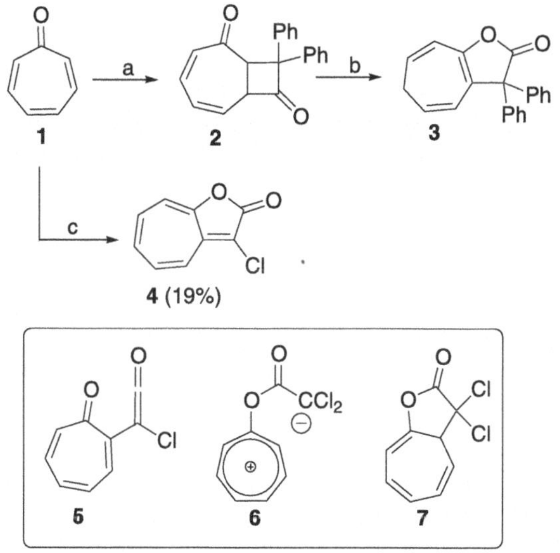

Scheme 2 Reagents and conditions: (a) $Ph_2CHCOCl$, Et_3N, ether, 0 °C; (b) quinoline/Δ; (c) $Cl_2CHCOCl$, Et_3N, ether, 0 °C

Machiguchi and co-workers have recently shown that the reaction of ketenes with tropone does not proceed by a concerted pathway [2]. Spectroscopic studies as well as theoretical investigations support a non-concerted pathway proceeding through a dioxetane intermediate **9** formed by [2+2] cycloaddition. The latter undergoes isomerization to the β-lactone intermediate **11** via a zwitterion, which again gets transformed to the [8+2] adduct via a second zwitterionic intermediate (Scheme 3).

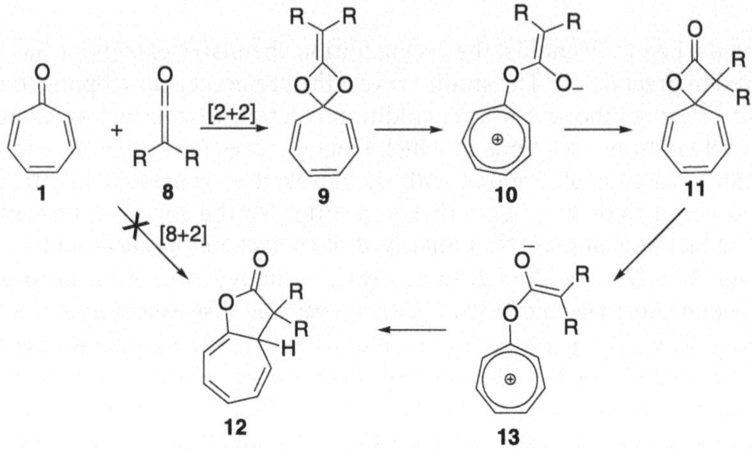

Scheme 3

Tropone was found to react with isocyanates resulting in the formation of corresponding imino derivatives after extrusion of CO_2 from the initial adducts [3]. The imino derivatives thus formed reacted with excess isocyanates to give the corresponding 1 : 2 adducts **15**. The product arises via an [8+2]-type annulation which may well involve a zwitterionic intermediate (Scheme 4).

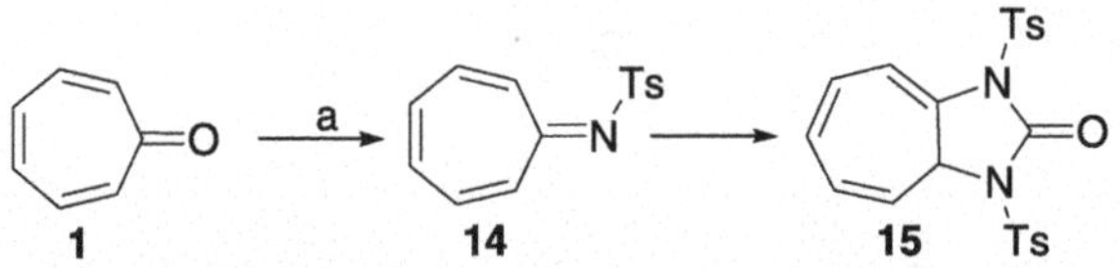

Scheme 4 Reagents and conditions: (a) Ts–N=C=O, CH_2Cl_2, –78 °C to rt

The reaction of tropone with enamines delivered different products depending on the substituents on the enamine [4]. Thus, 1-morpholinocyclohexene reacted with tropone to afford the [8+2]-type adduct **17** in 77% yield. 1-Morpholinocyclopentene also gave a similar adduct in good yield. The product may be considered to arise via an [8+2] addition or via a nucleophilic addition of the enamine to tropone and subsequent ring closure. On the other hand, 1-morpholinopropene afforded a [4+2]-type adduct as shown in Scheme 5 in 66% yield.

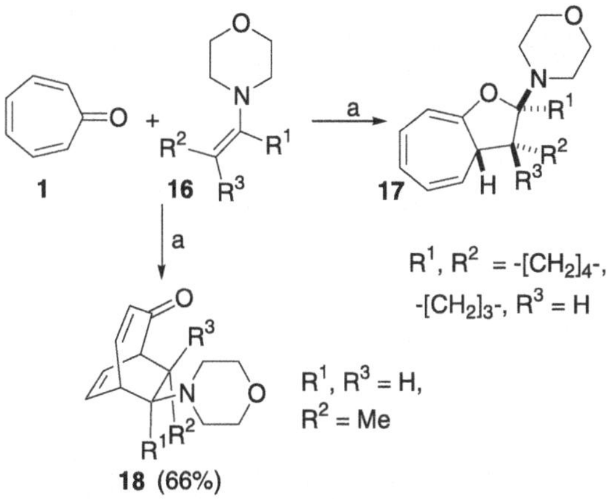

Scheme 5 Reagents and conditions: (a) neat (or EtOH/C_6H_6/THF), rt

Tropone has been shown to undergo photodimerization to yield a [4+2], [6+2] or [6+4] adduct [5,6]. Cantrell reported that tropone undergoes an [8+2] annulation with simple alkenes to yield the 8-oxabicyclo[5.3.0] ring system on irradiation (Scheme 6) [7]. The reaction is proposed to proceed through the tropone triplet species involving an $n \rightarrow \pi^*$ transition from the carbonyl oxygen to the alkene moiety. Product **20**, on heating underwent sigmatropic hydrogen shifts to afford the isomers **21** and **22**.

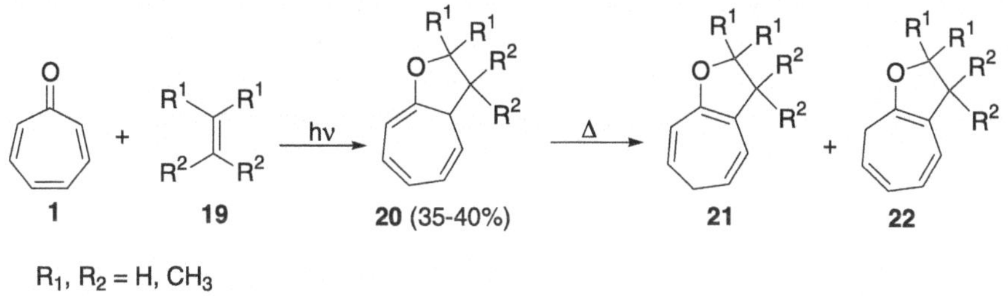

Scheme 6

Subsequently, the photochemical reaction of tropone was carried out with 1,1-dialkoxyethenes to afford the corresponding [8+2] adducts in around 40% yield (Scheme 7). Interestingly, the thermal reaction of tropone with 1,1-diethoxyethene afforded the [8+2] adduct **24** (60%) along with the [4+2] adduct **25** (22%) [8].

8-Oxoheptafulvene **26** reacted with tropones to give the adducts **27**, incorporating a norcaradiene ring system (Scheme 8) [9,10]. These are proposed to arise via [8+2] cycloaddition reactions of tropones **1a** to 8-oxoheptafulvene. Products **27** were found to undergo rearrangement to **28** on heating in the

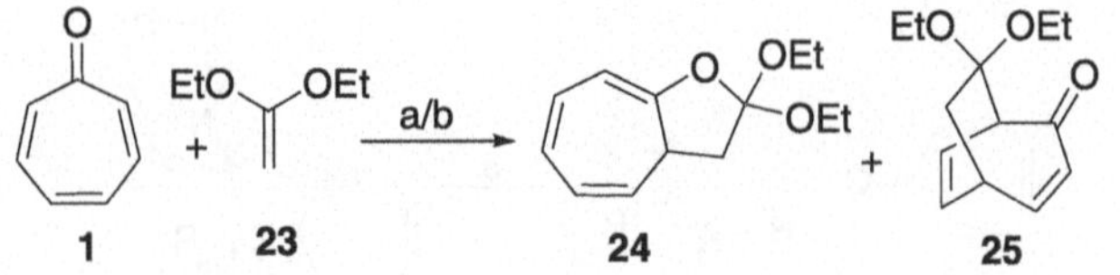

Scheme 7 Reagents and conditions: (a) hν, **24** (40%); (b) Δ, **24** (60%) + **25** (22%)

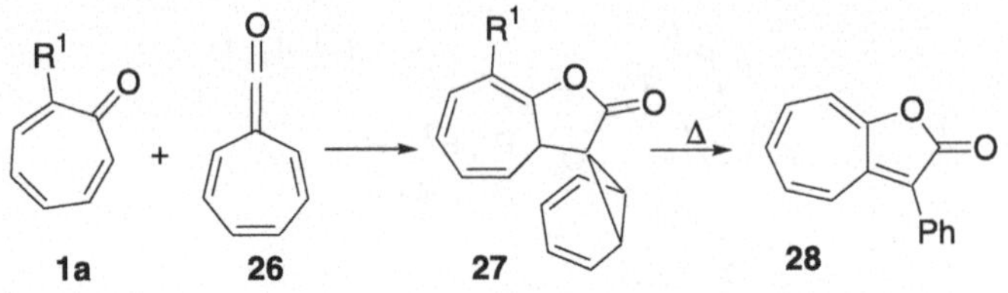

Scheme 8

presence of oxygen. Benzophenone, on the other hand, gave an [8+2] annulation product with 8-oxoheptafulvene which was proposed to arise from the corresponding [2+2] adduct via rearrangement.

Truce and Lin have reported the formation of 1 : 1 [8+2] adducts between tropone and sulfene [11]. A THF solution of triethylamine was added dropwise to a solution of tropone and α-toluene sulfonyl chloride in THF at 0 °C. Detailed analysis revealed that the adduct isolated was the *cis* isomer. On treatment with base, the primary adduct was transformed to the *trans* isomer. The initial exclusive formation of the *cis* adduct indicates that the reaction is either proceeding by a concerted route or that the zwitterionic intermediate collapses to the *cis* adduct before bond rotations. The adducts on heating lost SO_2 to form the corresponding *cis* or *trans* hydroxystilbenes or styrenes (Scheme 9).

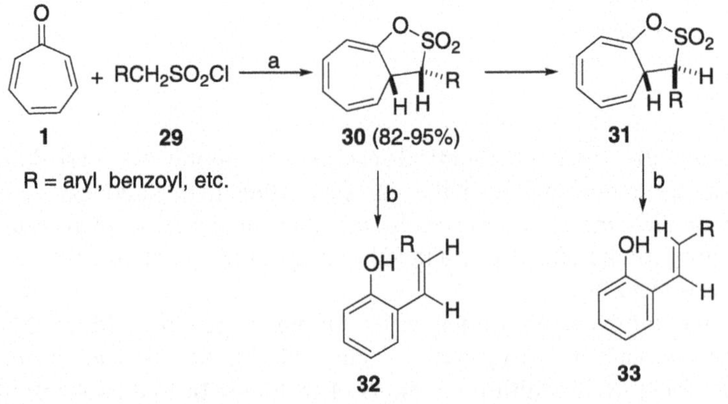

Scheme 9 Reagents and conditions: (a) Et_3N, THF, 0 °C, N_2; (b) Δ, $-SO_2$

Ciabattoni and Cabell have reported that the reaction of tropone with an excess of mesyl sulfene afforded the corresponding 1 : 2 adduct in 40% yield [12].

The phosphene **35** generated photolytically from the α-diazo phosphine oxide **34** underwent [8+2] cycloaddition with tropone [13]. The adduct on heating afforded **37** through a [1,5] sigmatropic hydrogen shift (Scheme 10).

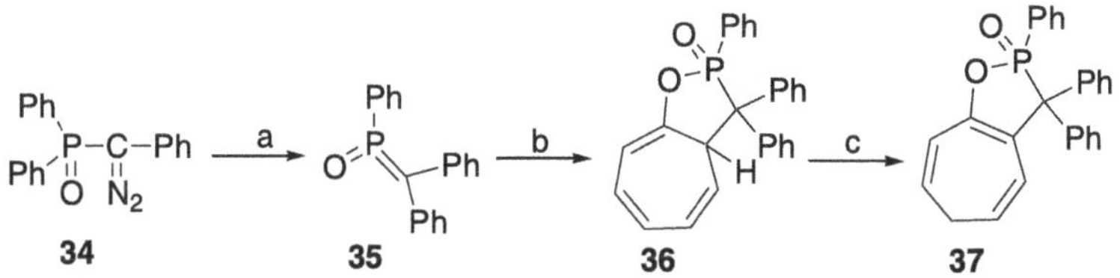

Scheme 10 Reagents and conditions: (a) benzene, hν, $-N_2$; (b) tropone; (c) 180 °C [1,5,6], H shift

Tropone was shown to undergo exclusive [8+2] cycloaddition with phenyl sulfonyl allene [14]. When the two were heated at 100 °C without solvent or in benzene, the [8+2] adduct **39** was isolated in 41% and 31% yields, respectively. In acetonitrile, the product **40** was also observed which is proposed to arise via sigmatropic hydrogen shifts from **39**. Presumably, the trace amount of acid in acetonitrile facilitates the reaction (Scheme 11).

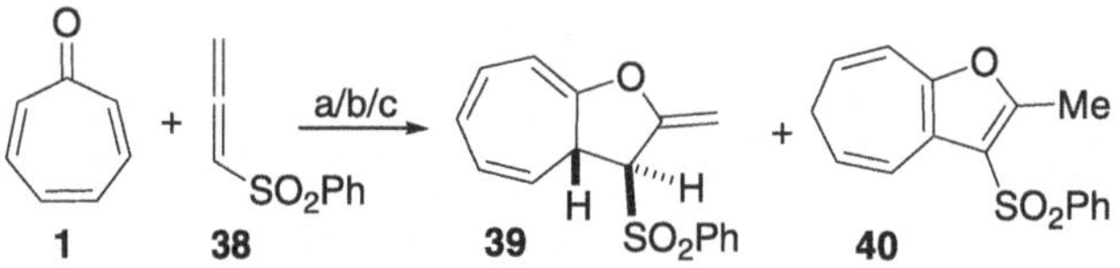

Scheme 11 Reagents and conditions: (a) CH_3CN, 100 °C, 16 h, **39** (32%) + **40** (10%); (b) benzene, sealed tube, 100 °C, 16 h, **39** (38%); (c) neat, 8 h, 100 °C, **39** (41%)

The reaction of tropone with phosphalkynes in toluene at 120 °C (5 days) led to the formation of cage compounds **45** [15]. The reaction proceeds through a series of [4+2], [2+2+2] and [8+2] addition reactions followed by rearrangements as depicted in Scheme 12. In the presence of excess alkyne at 95 °C, the reaction affords compound **42** in 52% yield.

Tropone reacted with naphthocyclopropenes via an [8π+2σ]-type pathway to yield cyclic ethers as shown in Scheme 13 [16]. The reaction was proposed to proceed through the intermediate **47**, which rearranges through path **a** or **b** as shown.

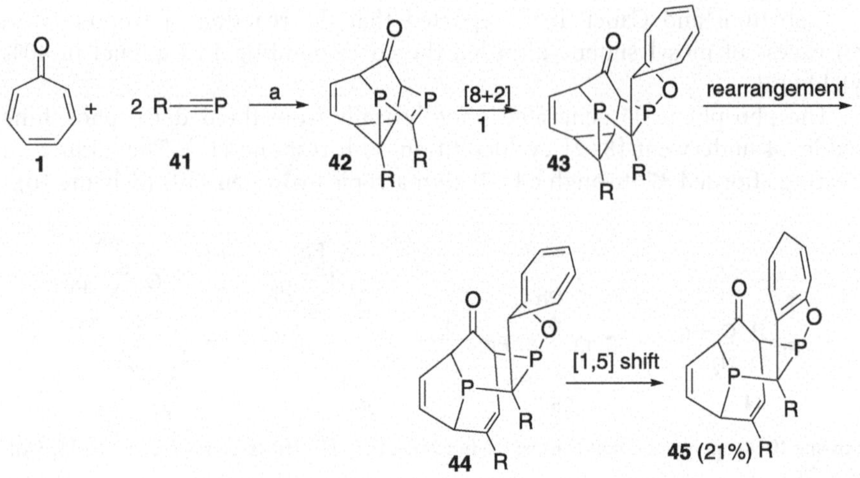

Scheme 12 Reagents and conditions: (a) toluene, 120 °C, Schlenk tube

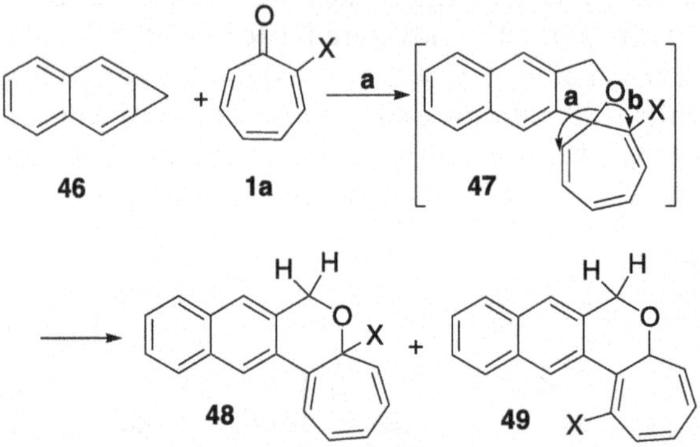

Scheme 13 Reagents and conditions: (a) $CHCl_3$, Yb(fod)$_3$

3
Reactions of Tropothione

Although tropothione was known in the literature from the early 1960s, the first report on its cycloaddition reaction was published only in 1973 by Machiguchi et al. [17]. A benzene solution of **50** reacted with maleic anhydride giving rise to the [8+2] cycloadduct **52** in quantitative yield. The reaction with DMAD (dimethyl acetylene dicarboxylate), on the other hand, provided the thiachromene derivative **56** in 65% yield. The latter is proposed to arise from the initial [8+2] adduct through rearrangement as depicted in Scheme 14.

The successful isolation of tropothione and a detailed investigation of its spectroscopic, physical as well as chemical properties were reported by

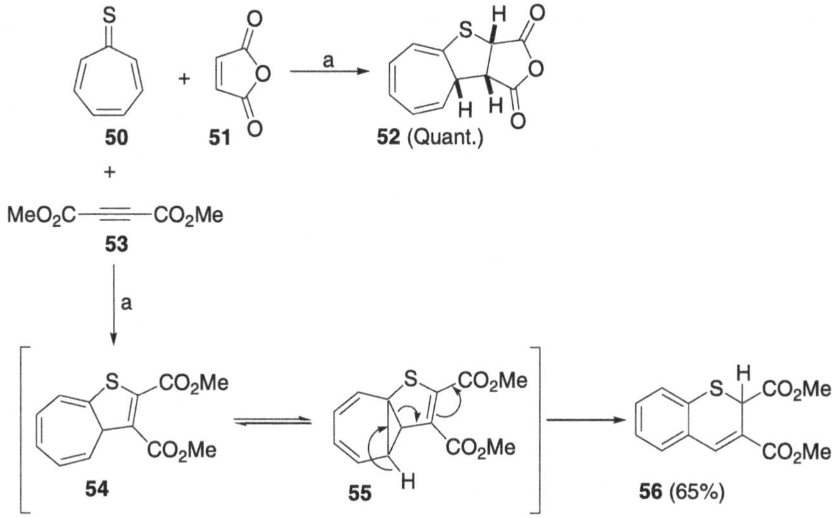

Scheme 14 Reagents and conditions: (a) benzene, 10 °C, 30 min

Machiguchi et al. [18–21]. The molecule was found to be stable at –78 °C under nitrogen, but dimerized readily at 0 °C. A 1 molar solution of **50** was found to have a half life of 9.1 h at 25 °C. On oxidation tropothione gave tropone whereas, on reaction with hydrazine and hydroxylamine, tropone hydrazone and tropone oxime respectively were isolated in good yield. These studies revealed that the structure of tropothione was closer to **50a** than to **50b**, with less polar and less basic properties compared to tropone.

The dimerization of tropothione afforded an [8+8]-type dimer **57** of the head-to-tail type. The reaction represents the first example of a non-topochemically controlled reaction in the ground state (Scheme 15) [22].

In contrast to tropone and azaheptafulvenes, tropothione underwent an efficient [8+2] cycloaddition reaction with cyclopentadiene [23]. The endo

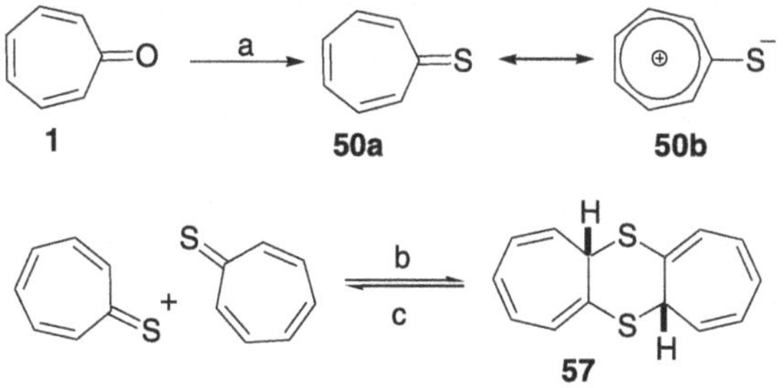

Scheme 15 Reagents and conditions: (a) P_2S_5, Et_3N, CH_2Cl_2, N_2, 0 °C, 94%; (b) crystalline state, 0 °C, 2 days, 98%; (c) rt in solution or at mp in the molten state

cycloadduct **59** was isolated in 89% yield, but underwent isomerization through sigmatropic hydrogen shifts at higher temperature or in the presence of acid. Both adducts were oxidized to the corresponding thermally stable sulfones with monoperphthalic acid to aid structural elucidation (Scheme 16). Kinetic studies supported a concerted reaction pathway proceeding through a highly oriented 10π transition state.

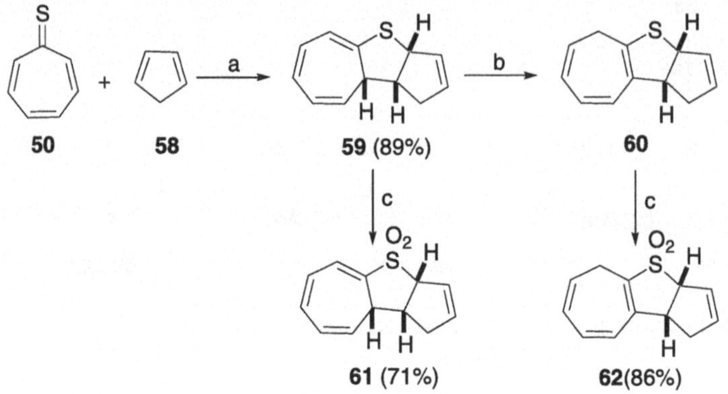

Scheme 16 Reagents and conditions: (a) $CHCl_3$, 40 °C; (b) toluene, reflux or, CF_3COOH (cat.), rt; (c) monoperphthalic acid, $CHCl_3$, –6 °C

Tropothione reacted with chloroketene, generated in situ, in benzene at room temperature to afford the 1 : 2 adduct **66** in 68% yield [24]. The formation of **66** is rationalized by invoking an electrophilic addition of chloroketene to the initially formed adduct as shown in Scheme 17. The controlled addition of chloroacetyl chloride to a solution of tropothione in CH_2Cl_2 containing Et_3N allowed separation of the 1 : 1 [8+2] adduct **65** in 72% yield. An MO calculation revealed that the HOMO of **65** has the highest lobe on C-3 which is the preferred site for attack by the electrophile.

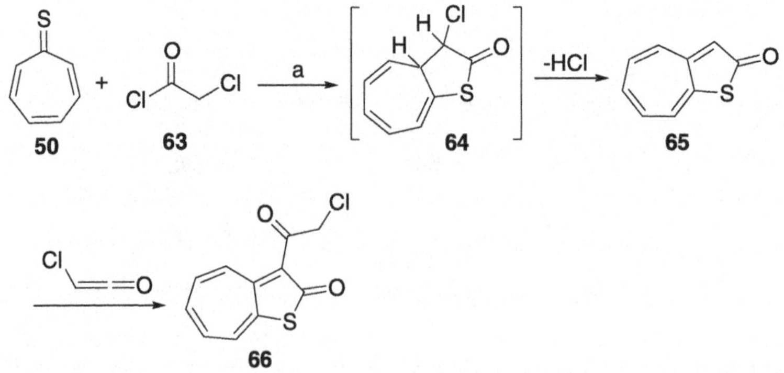

Scheme 17 Reagents and conditions: (a) Et_3N, benzene, rt

A very low yielding synthesis of the 2*H*-cyclohepta[b]thiophene system **69** was reported by Pietra and co-workers [25]. Tropothione **50** was generated in situ from tropone by chlorination with oxalyl chloride in dichloromethane followed by treatment with hydrogen sulfide in the presence of potassium carbonate. To a solution of **50** in dichloromethane was added dichloroacetyl chloride followed by triethylamine. The thus-generated dichloroketene apparently underwent [8+2] cycloaddition followed by dehydrochlorination to give **69**. It is quite likely that the reaction is initiated by the nucleophilic attack of tropothione sulfur on the ketene (Scheme 18).

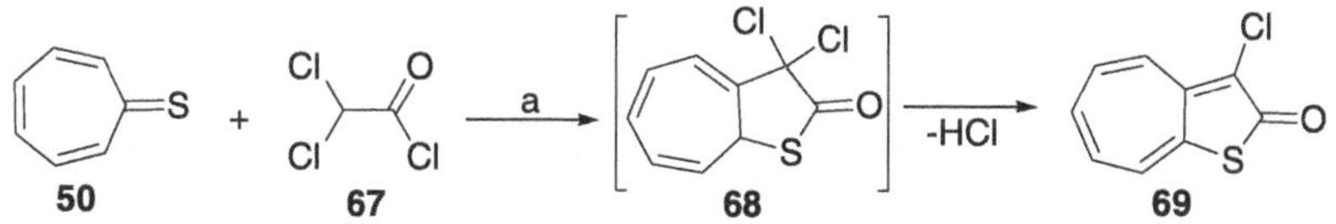

Scheme 18 Reagents and conditions: (a) Et_3N, CH_2Cl_2, N_2, rt

Diphenyl ketene was found to undergo an efficient [8+2] cycloaddition reaction with tropothione (Scheme 19) [26]. An MO calculation was performed in order to throw light on the orientation of the [8+2] cycloaddition, which revealed an orbital mixing of the phenyl group with two vacant FMOs of the ketene moiety. This provides an appropriate site for the antisymmetric orbital interaction with the tropothione molecule.

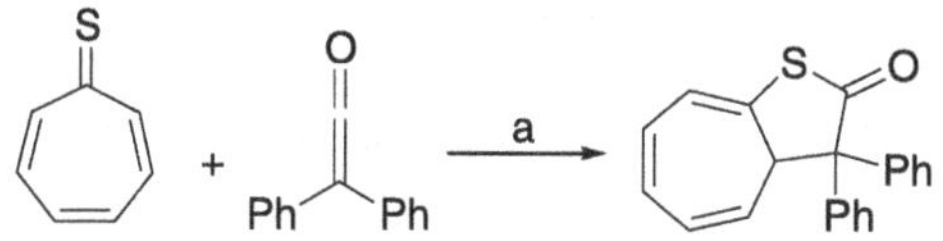

Scheme 19 Reagents and conditions: (a) benzene, 10 °C

Machiguchi's group carried out a theoretical investigation of two model reactions involving ethylene to butadiene and to tropothione to compare the reactivities in [4+2] and [8+2] cycloaddition reactions [27]. The results indicated that both the [8+2] and [4+2] cycloadditions proceed with small activation energies. The high reactivity is ascribed to the larger lobe of the FMO (HOMO or LUMO) on the sulfur atom. Also, the Diels–Alder reaction was found to be of the inverse electron demand type. The propensity for [8+2] cycloaddition was slightly lower owing to the delocalization of the FMOs in the seven-membered ring.

Machiguchi et al. also reported a comparative study of the reactivities of tropone and tropothione towards pentafulvenes [28]. Tropone was reported to react with dimethyl fulvene to afford a double [6+4] adduct via an initially formed [6+4] monoadduct. Tropothione, on the other hand, underwent an exclusive [8+2] cycloaddition reaction with dimethyl- as well as diphenyl pentafulvene (Scheme 20). A kinetic study of the tropothione-dimethyl fulvene reaction indicated that it proceeds via a concerted endo $[{}_{\pi}8_S + {}_{\pi}2_S]$ route. To explain the [8+2]/[6+4] and the *endo/exo* selectivities observed in the tropothione reaction, a theoretical investigation was also performed. The results indicate that the large lobe of the HOMO on the sulfur atom is responsible for the selective [8+2] cycloaddition whereas, the secondary orbital interactions at the initial stage of the intrinsic reaction coordinate gives rise to the *endo* configuration.

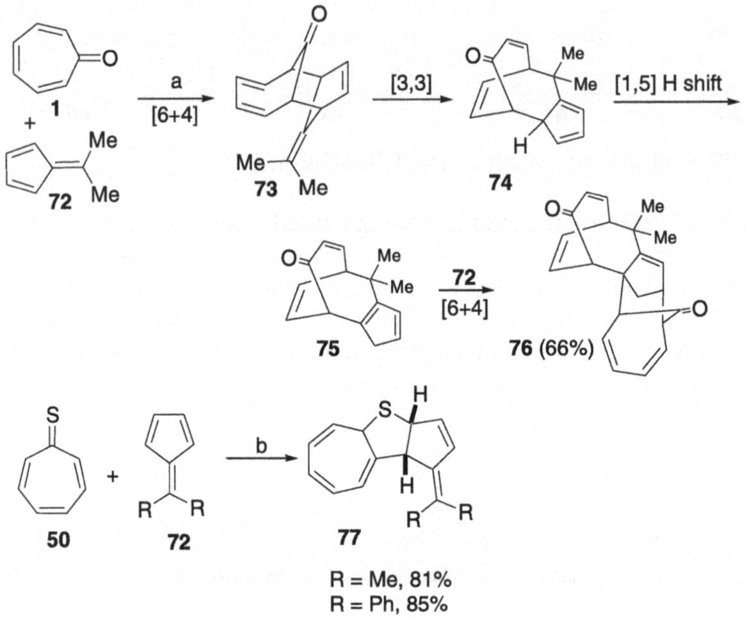

Scheme 20 Reagents and conditions: (a) THF, 50 °C; (b) benzene, 10 °C

The synthesis of tropothione-*S*-oxide, the first example of a sulfine charge reversion (umpolung) has been reported [29]. The synthesis is achieved by careful oxidation of tropothione with *m*-CPBA below –60 °C (Scheme 21). The ring is found to be positively polarized which is in contrast to the parent sulfine ($CH_2{=}S{=}O$), where the carbon acquires a partial negative charge. The molecule does not react with DMAD which is a representative electron-deficient alkyne, but reacts readily with an electron-rich enamine. This is in agreement with the predicted electron deficiency of the triene system.

In an interesting extension to this work, Machiguchi et al. have reported the in situ generation of tropothione-*S*-sulfide [30]. The molecule was generated by treatment of tropone hydrazone with disulfur dichloride in deoxygenated chloroform at –78 °C. On treatment with DMAD the molecule was

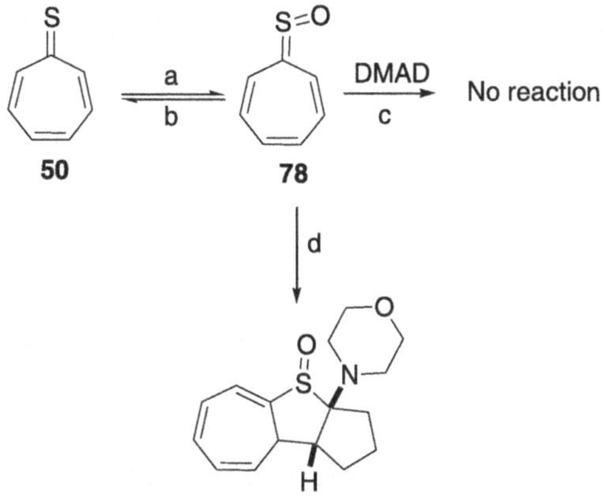

Scheme 21 Reagents and conditions: (a) *m*-CPBA, CH_2Cl_2, –60 °C, 2 h, 75%; (b) P_4S_{10}, CH_2Cl_2, –5 °C, 1.5 h, 91%; (c) $CHCl_3$, reflux, 2 days; (d) *N*-cyclopentenomorpholine, CH_2Cl_2, rt, 4 h

trapped in a novel [10+2]-type cycloaddition. The reaction is proposed to proceed through the sequence outlined in Scheme 22. Theoretical studies also support the proposed pathway involving a zwitterionic intermediate.

The same group has reported a comparative study of the cycloaddition reactions of maleic anhydride with tropone and tropothione [31]. With tropone the [4+2] adduct was obtained in 70% yield, whereas tropothione gave the [8+2] adduct in 98% yield (Scheme 23). Ab initio calculations on the transition states and products of the two reactions revealed that the [4+2] *endo* path is favorable both thermodynamically and kinetically for the tropone addition

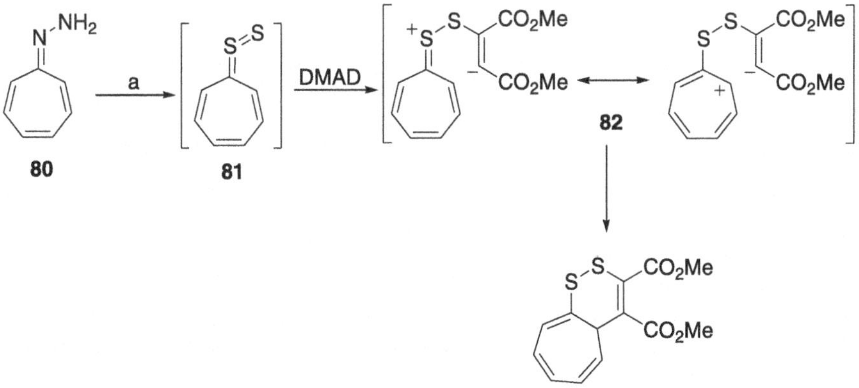

Scheme 22 Reagents and conditions: (a) S_2Cl_2, Et_3N, $CHCl_3$, –78 °C

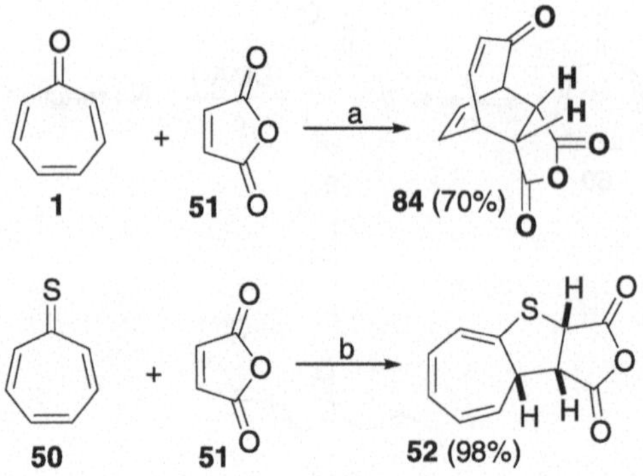

Scheme 23 Reagents and conditions: (a) xylene, reflux, 12 h; (b) $CHCl_3$, 0 °C, 30 min

reaction. On the other hand, the [8+2] *endo* path is favored kinetically for the tropothione reaction.

Hafner and co-workers have described the synthesis of pentafulvenyl tropone **86** and, in turn, its attempted conversion to the pentafulvenyl tropothione **87** (Scheme 24). Presumably the latter was formed but it underwent an interesting 10π electrocyclization to afford the spirocompound **88** in 49% yield [32].

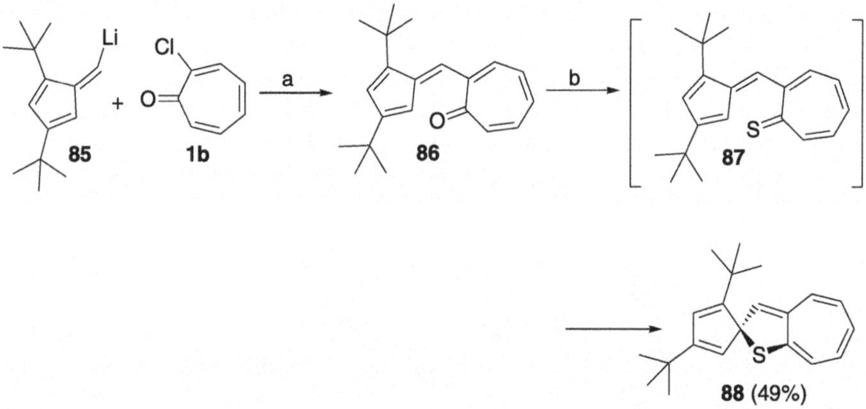

Scheme 24 Reagents and conditions: (a) THF/Et_2O/*n*-pentane (4:1:1), –110 °C, 71%; (b) P_2S_5, Et_3N, 0 °C

Recently, Nair et al. have reported the utilization of the Lawesson reagent for the in situ generation of tropothione and novel [8+2] cycloaddition reactions of the latter with a range of dienophiles, for which an example is shown in Scheme 25 [33].

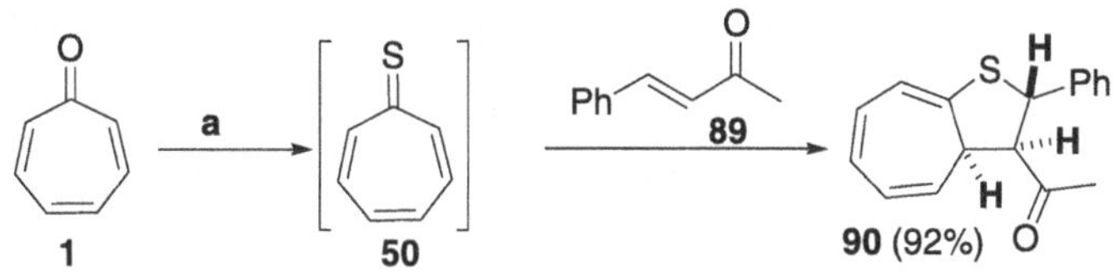

Scheme 25 Reagents and conditions: (a) Lawesson's reagent, benzene, rt

4
Reactions of Azaheptafulvenes

Cycloaddition reactions of 8-alkyl- and 8-aryl-8-azaheptafulvenes with isocyanates and isothiocyanates were reported by Kanemasa and co-workers [34] (see also Scheme 4 above). Isocyanates reacted as 2π components through the N=C bond to afford the corresponding [8+2] adducts in good to excellent yields (Scheme 26). Isothiocyanates, on the other hand, reacted through the C=S bond to afford the corresponding imino cycloheptathiazoles **94**. Benzoyl and thiobenzoyl isocyanates followed a reaction course analogous to those of isocyanates. Interestingly, benzoyl isothiocyanate yielded the corresponding [8+4] adducts **93** in excellent yields. In order to explain the formation of different kinds of products, the intermediacy of zwitterions such as **95** was invoked (Scheme 26).

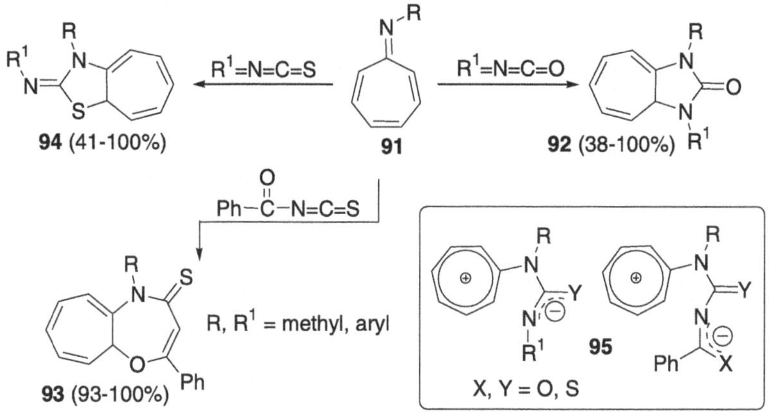

Scheme 26

On extending the reaction of sulfenes with tropone (see Scheme 9) to azaheptafulvenes, Truce and Shepherd observed that the corresponding γ-sultams (Scheme 27) were formed via analogous [8+2] cycloaddition reactions [35]. Interestingly, alkyl sulfenes as well as the parent sulfene underwent efficient [8+2] addition reactions. Here again as with tropone, the *cis* adducts were the initial products which isomerized to the *trans* isomers on treat-

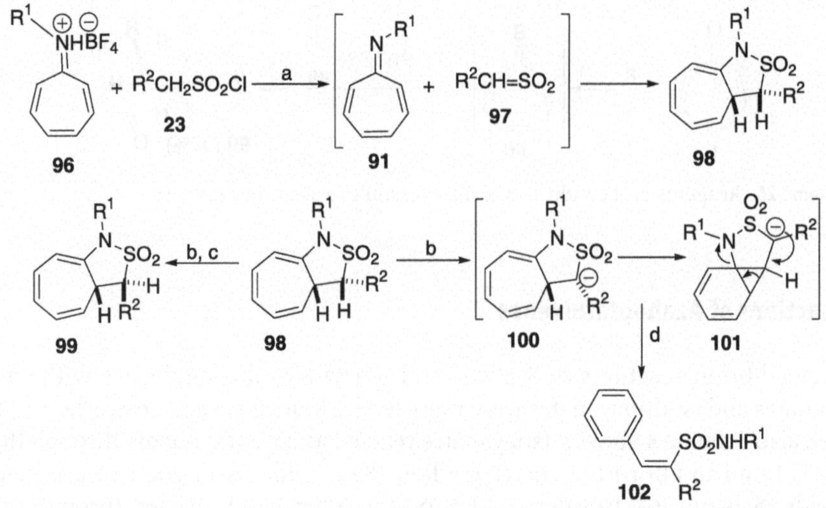

Scheme 27 Reagents and conditions: (a) Et_3N, THF, 0 °C; (b) *n*-BuLi, THF, –78 °C; (c) H_2O, –78 °C; (d) H_2O, 25 °C

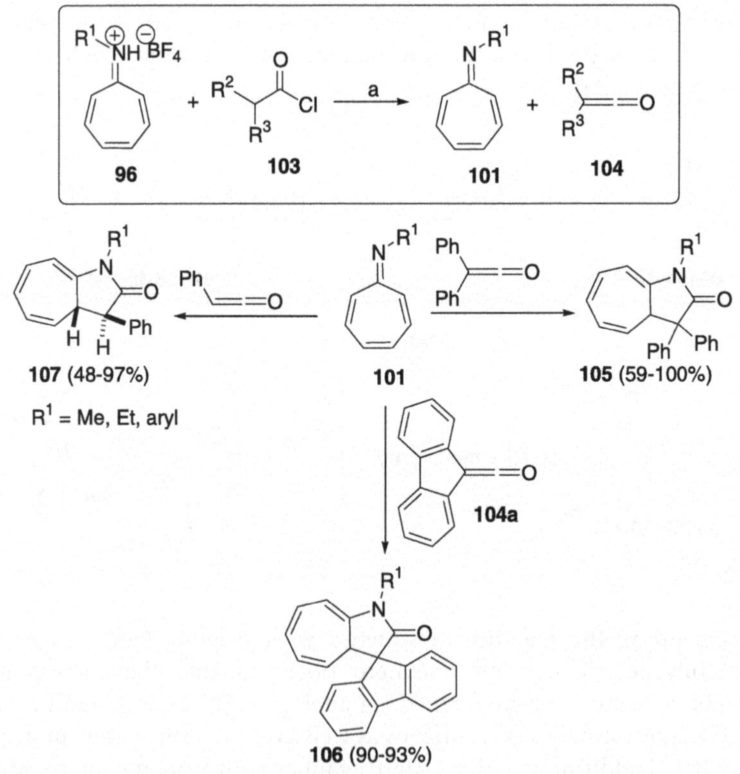

Scheme 28 Reagents and conditions: (a) Et_3N, CH_2Cl_2, rt

ment with *n*-BuLi at −78 °C in THF. When the solution of sultam anion **100** was warmed to room temperature without quenching, an interesting rearrangement took place which was proposed to involve a norcaradiene-type intermediate, and the products isolated were the *ω*-styrene sulfonamides **102**. In contrast to the *γ*-sultones, the adducts did not eliminate SO_2 even at 145 °C.

Kanemasa and co-workers have reported the addition of ketenes to 8-aryl-8-azaheptafulvenes, with formation of [8+2] adducts in good yields [36, 37]. The azafulvenes as well as the ketenes were generated in situ in dichloromethane solution with the aid of triethylamine as the base (Scheme 28). A monosubstituted ketene, viz. phenyl ketene, exclusively gave the *trans* [8+2] adduct, and diphenyl ketene as well as **104a** also gave the corresponding [8+2] cycloadducts in excellent yields.

8-(*p*-Chlorophenyl)-8-azaheptafulvene **91** underwent a cycloaddition reaction with DMAD to afford three different products as shown in Scheme 29 [38]. The initial product was the [8+2] adduct **108**, which rearranged through hydrogen shift or ring contraction to yield **109**, **110** and **111**. The product ratio was found to be highly dependent on the polarity of the solvent. In a polar solvent like methanol **109** was the major product, whereas in benzene **111** was the major product; the proportion of **110** remained almost the same.

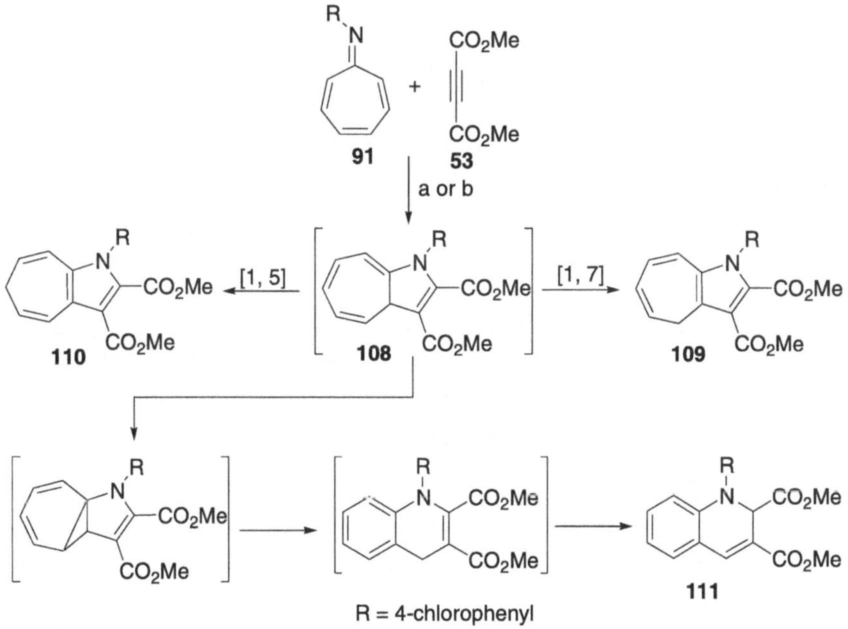

Scheme 29 Reagents and conditions: (a) methanol, 40–50 °C, 0.5 h, 80% (**109**:**110**:**111** = 75:25:0); (b) benzene, rt, 4 days, 76% (**109**:**110**:**111** = 0:31:69)

Presumably product **109** is formed via a [1,7] hydrogen migration whereas, the azachromene formation occurs via the norcaradiene-type intermediate as depicted in Scheme 29.

Gandolfi and Toma have reported the cycloaddition reactions of diphenyl nitrilimine **112** with azaheptafulvenes yielding the corresponding [8+4] adducts (Scheme 30) [39]. The reaction was proposed to proceed via an initial [3+2] adduct **113,** which rearranges to the observed products. Even though the intermediate could not be detected in the above reaction, the corresponding iron carbonyl complexed azafulvenes reacted to give the [3+2] adducts as stable compounds. The latter on decomplexation with trimethylamine oxide led to the formation of the same adducts **114** and **115**, thus lending some support to the proposed mechanistic pathway.

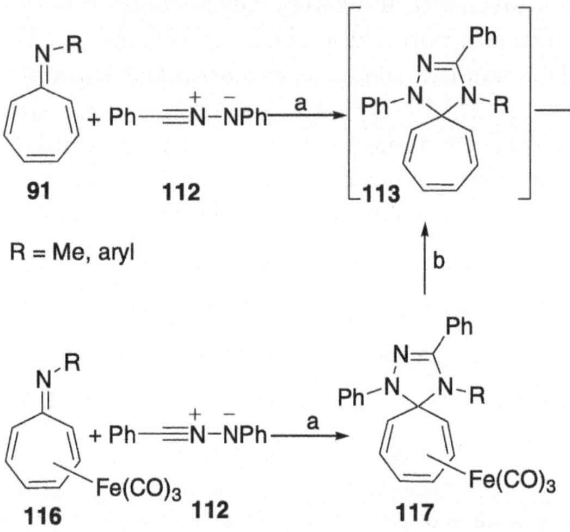

Scheme 30 Reagents and conditions: (a) CH_3CN, rt; (b) $(CH_3)_3N^+O^-$, CH_3CN, rt

Azaheptafulvenes **91** reacted with phenyl sulfonyl allene to yield the [8+2] adducts **118** (Scheme 31) which isomerized gradually at room temperature to yield the stable crystalline products **120** in 80–90% yields [40].

The reaction rate was considerably enhanced by polar solvents suggesting that the reaction may proceed through a polar transition state or a zwitterionic intermediate like **121** (Scheme 31).

The reaction of azaheptafulvenes with carbondisulfide has been reported to yield the corresponding [8+2] adducts quantitatively (Scheme 32) [41, 42]. The initially formed adducts **123** rearranged through [1,5] sigmatropic hydrogen shifts to **124**.

The reaction of 8-aryl-8-azaheptafulvenes with 8-oxoheptafulvene proceeded through a [8+2] cycloaddition pathway to form the corresponding dihydroazazulanones possessing spirocyclic and norcaradiene moieties

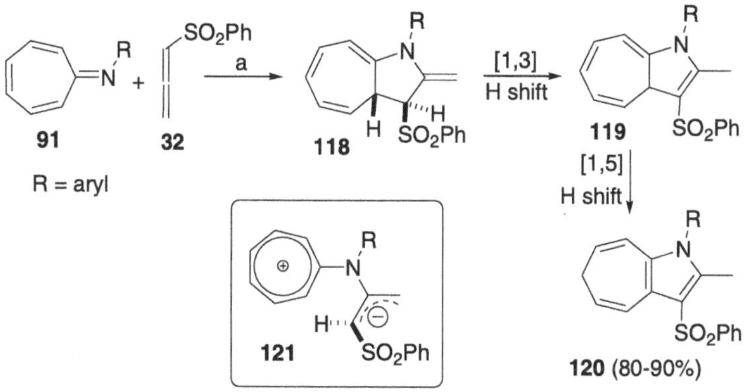

Scheme 31

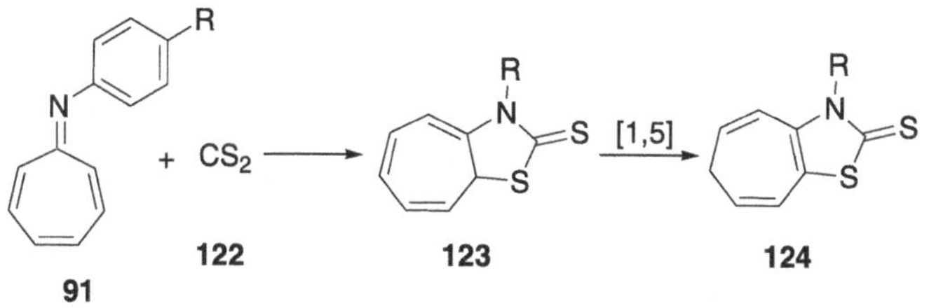

Scheme 32

(Scheme 33) [43]. The phenyl and tosyl hydrazones of tropone also gave a similar reaction.

Gandolfi and co-workers have reported the cycloaddition reactions of azaheptafulvenes with cyclopentadienones (Scheme 34) [44]. In analogy with tropone, the *exo* [6+4] adducts were the major products along with minor amounts of *endo* [4+2] adducts and trace amounts of [8+2] adducts. It was observed that the [6+4] adducts at higher temperatures underwent [3,3] sigmatropic rearrangement to afford the [8+2]-type adducts. It was shown by control experiments that the [8+2] adducts were not formed by cycloreversion of the initial [6+4] adduct followed by [8+2] cycloaddition.

1-Azaazulenes (e.g. **132**) are attractive targets in organic synthesis [45–49]. In analogy with the enamine method for the synthesis of azulenoid com-

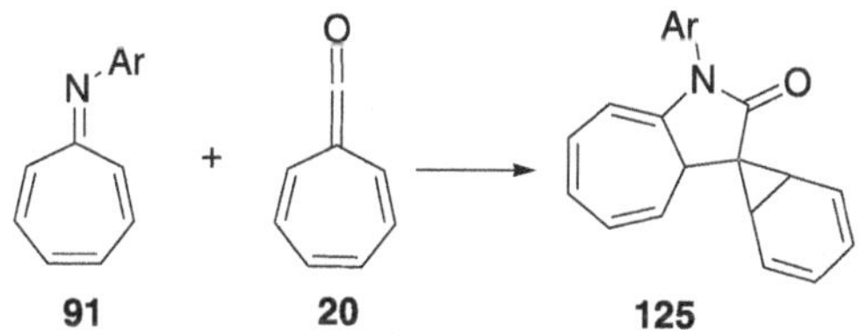

Scheme 33

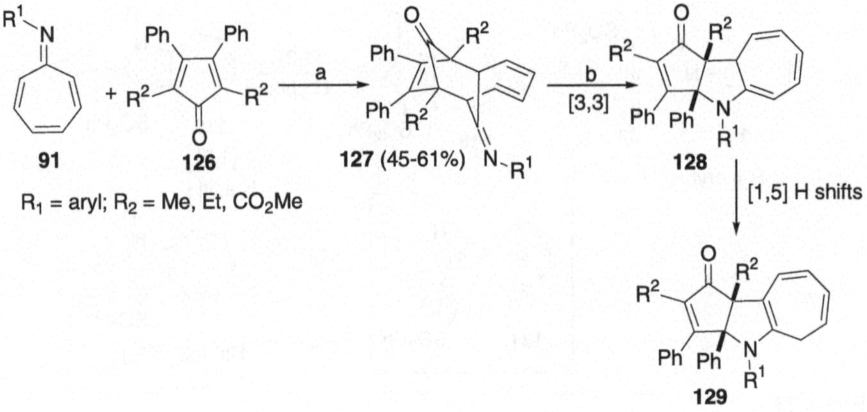

Scheme 34 Reagents and conditions: (a) benzene, reflux, N_2; (b) toluene, reflux, N_2

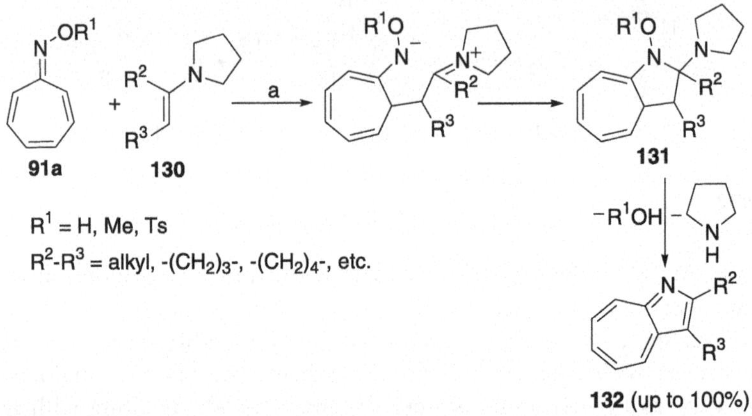

Scheme 35 Reagents and conditions: (a) benzene, reflux

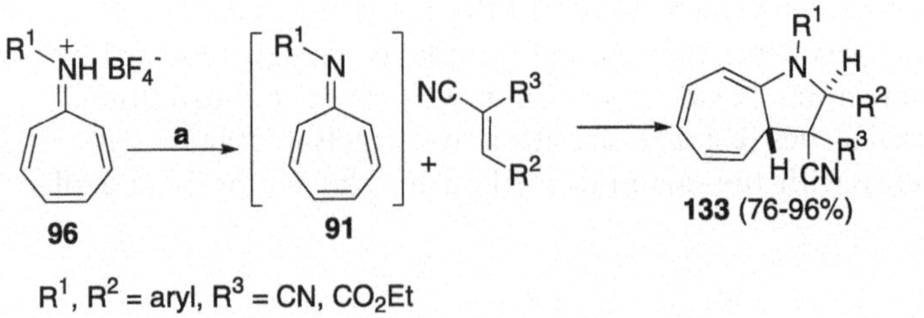

Scheme 36 Reagents and conditions: (a) Et_3N, CH_2Cl_2, rt

pounds [50–62], Takayasu and Nitta reported the synthesis of novel 1-aza-azulene derivatives through cycloaddition reactions of azaheptafulvenes with enamines (Scheme 35) [63]. Azaazulene derivatives with a leaving group on the *N*-atom reacted with enamines to afford the corresponding azaazulenes after elimination of the leaving group. Azaazulenes annulated with heterocycles were also synthesized by employing enamines derived from 3- and 4-piperidones and 3-oxotetrahydrothiophene.

In contrast to tropothione, azaheptafulvenes did not undergo any [8+2] cycloaddition with cyclopentadiene or pentafulvenes. The [4+2] and/or the [6+4] adducts were obtained depending on the conditions [64].

Efficient cycloaddition reactions of azaheptafulvenes with activated styrenes were reported recently. The reactions afforded the corresponding dihydro-1-azaazulene derivatives **133** in high yields (Scheme 36) [65].

5
Miscellaneous Substrates

The utilization of the [8+2] cycloaddition strategy for the synthesis of oxa-bridged macrocycles was reported by Herndon and co-workers (Scheme 37) [66]. The dienyl isobenzofurans generated via coupling of the Fischer carbene complexes **135** with alkynes **134** underwent efficient addition reaction with DMAD, leading to the corresponding [8+2] adducts **137** along with minor amounts of the Diels–Alder products. Interestingly, the 11-oxa-bicyclo[6.2.1]undecane ring system, as in **137**, is a key structural feature of a group of compounds known as cladiellanes, many of which possess anti-inflammatory and anti-tumor activity.

Subsequently, the strategy was extended to simple dienyl furans, which lack the activation of a benzofuran (Scheme 38) [67]. The *Z*-dienyl furans re-

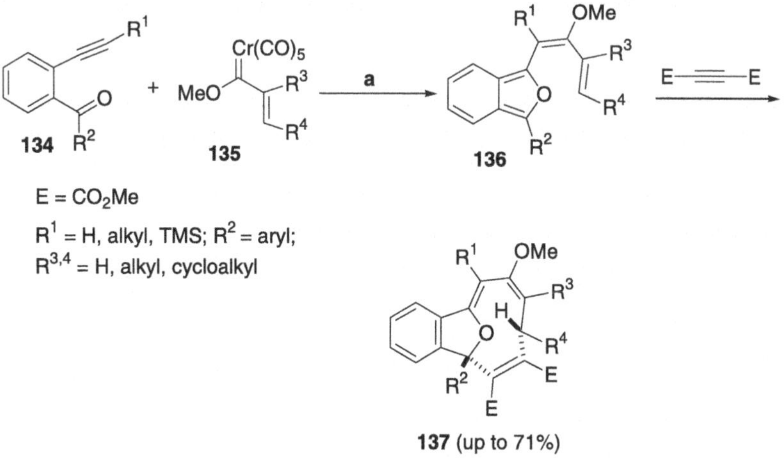

Scheme 37 Reagents and conditions: (a) dioxane, 85 °C, 1–2 h

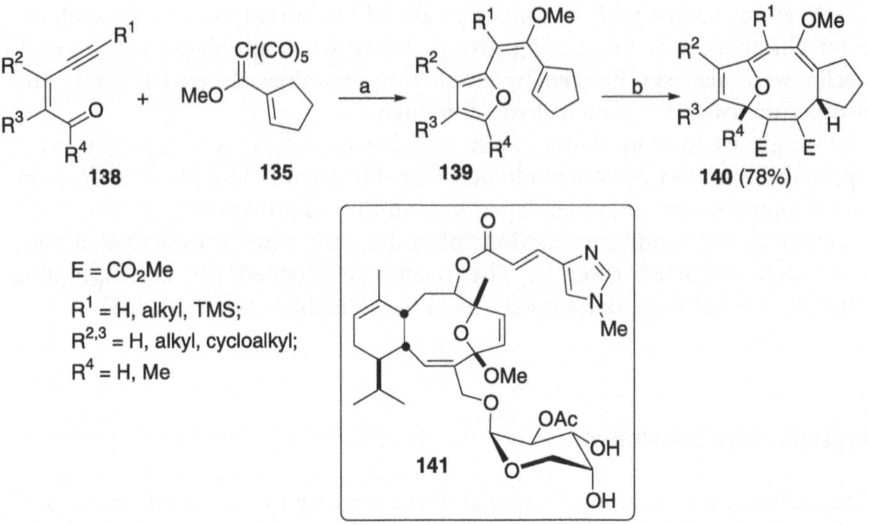

Scheme 38 Reagents and conditions: (a) dioxane, 80 °C; (b) DMAD, dioxane, 80 °C

quired were prepared by applying the Fischer carbene-mediated strategy on the enyne-carbonyl compounds **138**. The benzo-free macrocycle **140** presents a direct entry to the core structure of eleutherobin **141**, a potent anticancer agent.

The cycloaddition reactions of indolizines (e.g. **142**) with DMAD have been widely investigated. Godfrey observed that 1,2,6,7-dibenzoindolizine reacted with DMAD to afford the corresponding 1 : 1 [8+2] adduct [68]. Later, the reaction was extended to the parent indolizine and other derivatives by Boekelheide and co-workers [69–72]. The initial adducts on hydrolysis followed by decarboxylation afforded the corresponding cyclazine derivatives. The reaction was also extended to cyanoindolizines, which in some cases, afforded a 1 : 2 adduct as well [73].

Kanemasa and co-workers have reported the reaction of indolizines **142** with maleates and acrylates affording the corresponding [8+2] adducts **144** [74]. With maleic anhydride, the corresponding Michael adduct was obtained. *N*-Aryl maleimides afforded the [8+2] adducts along with the Michael addition products (Scheme 39). On the basis of these observations, a stepwise mechanism was proposed for this addition reaction.

Yerxa and Moore have reported the synthesis of indolizine diones **149** from cyclobutene diones **147** [75]. When the reaction was carried out in the presence of DMAD, the indolizine precursors **148** were trapped in [8+2] cycloaddition reactions to afford the pyrroloindolizines (cyclazines) **151** in very good yields (Scheme 40).

Jug and co-workers have carried out a theoretical investigation of the various mechanistic possibilities for the indolizine [8+2] cycloaddition [76].

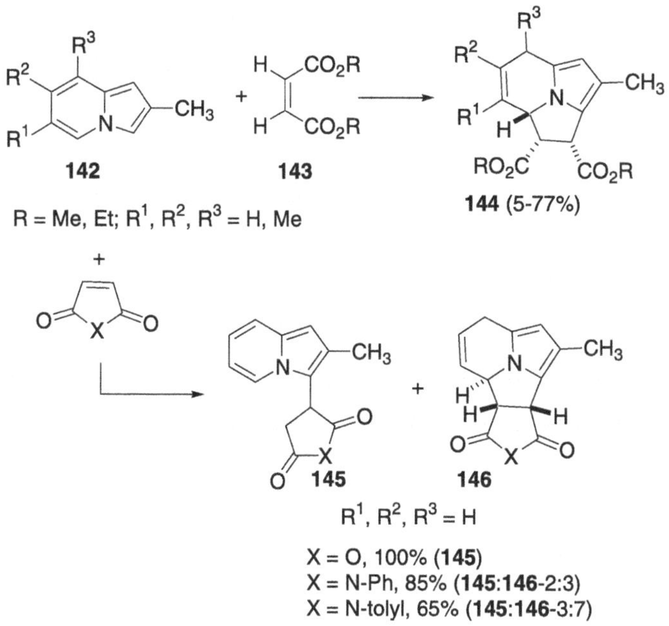

Scheme 39

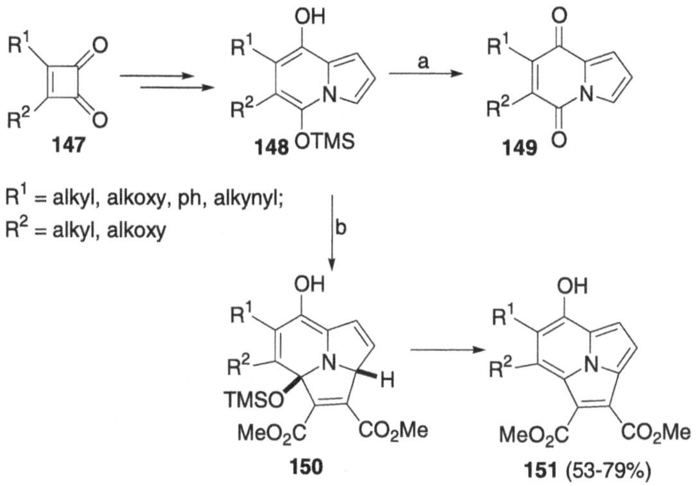

Scheme 40 Reagents and conditions: (a) air or $FeCl_3$; (b) DMAD, toluene, reflux

Indolizine and 6-nitroindolizine were the selected substrates and their reactions with ethylene, nitroethylene, methyl acrylate and dimethylamino ethylene were investigated. The results suggest a concerted mechanism except for the nitroethylene reactions. The reaction of dimethylamino ethylene with 6-

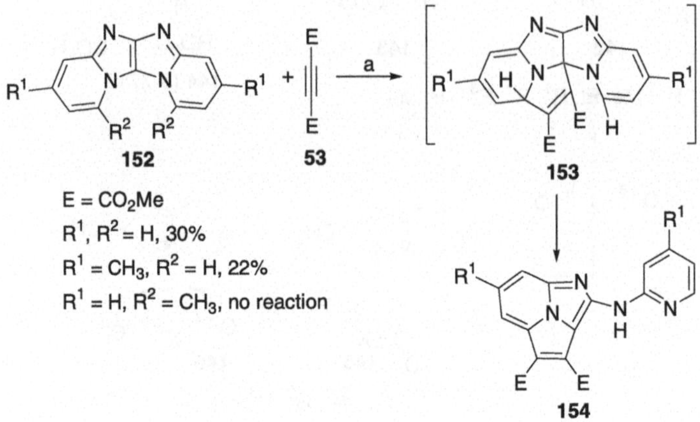

Scheme 41 Reagents and conditions: (a) methanol, reflux, HBr (cat.)

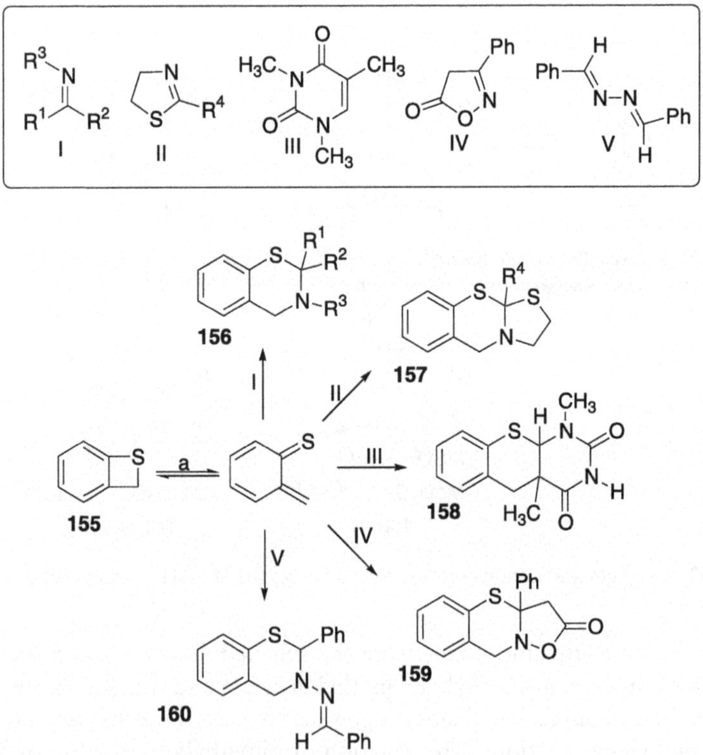

Scheme 42 Reagents and conditions: (a) toluene, reflux

nitroindolizine follows an inverse dipolar addition mechanism (nucleophilic addition followed by ring closure).

Matsumoto and co-workers have described the reaction of diazapentalenes with DMAD leading to the formation of unusual anti-Bredt's compounds [77]. The cycloaddition reactions of the tetraazapentalenes **152** with DMAD were reported by Leonard and Pereira (Scheme 41) [78]. The initial adduct was found to undergo ring cleavage to afford the cyclazines **154** as the final products. The ring-opening reaction was found to be accelerated by a catalytic amount of acid. Interestingly, the dimethyl compound **152** (R^2 = Me) which does not have hydrogen to migrate was found to be unreactive toward DMAD.

Benzothiete **155** undergoes thermal opening of the four-membered ring to form an 8π system which can undergo efficient cycloaddition reactions. Meier and co-workers have described the [8+2] cycloaddition reactions of **155** with a range of compounds containing C=N bonds [79]. Thus, compounds like imines, oximes, isoxazolinones, thiazolines, etc. reacted with **155** to form the corresponding [8+2] adducts in very low to good yields (Scheme 42).

Meier and Mayer have reported the synthesis of benzo bisthiete **162** which reacted with DMAD to afford the bis-adduct in 68% yield [80]. The reaction was extended to the synthesis of extended molecular frameworks. Thus, on reaction with the 2π component **161**, benzo bisthiete gave rise to the bis-adducts **163** in good yields [81]. These on acid treatment were transformed

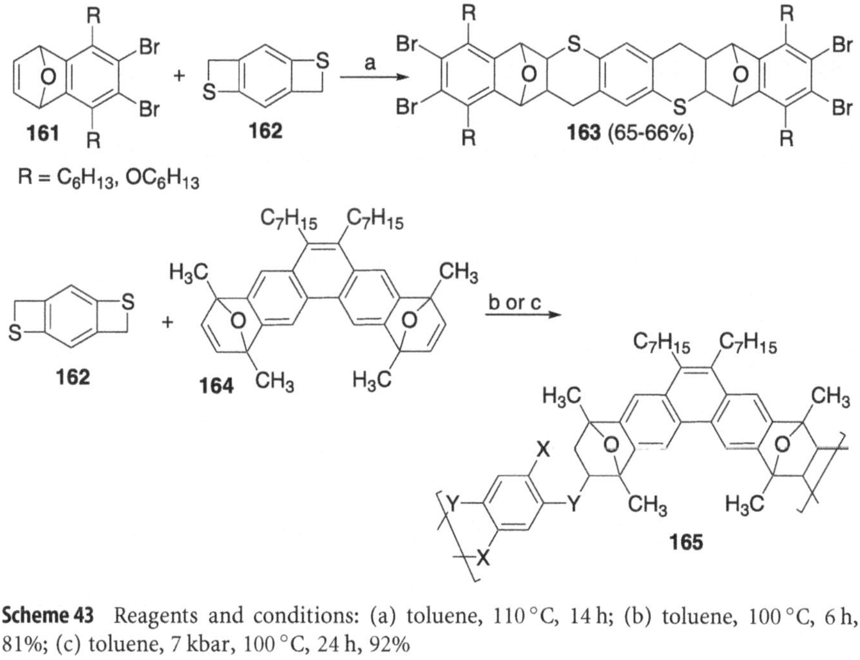

Scheme 43 Reagents and conditions: (a) toluene, 110 °C, 14 h; (b) toluene, 100 °C, 6 h, 81%; (c) toluene, 7 kbar, 100 °C, 24 h, 92%

to the aromatized products. The bis dienophile **164** was designed in order to effect multiple cycloadditions which can lead to bent-shaped polymers. Thus, the reaction of **162** with **164** afforded the polymeric compound **165** in good yield (Scheme 43).

6 Conclusion

In conclusion, it can be seen that [8+2] cycloaddition reactions of hetero-analogs of heptafulvene provide a facile entry into (7,5)-fused heterocyclic systems. The cycloaddition reactions of azaheptafulvenes and tropothione are especially noteworthy for their efficiency and very high yields of products. The [8+2] cycloaddition reactions of indolizines have generated a variety of cyclazine derivatives. The participation of dienylisobenzofurans in higher-order cycloadditions provides easy access to interesting macrocyclic systems with important biological activities. The reactions of benzothiete and benzobisthiete can be effectively utilized for the synthesis of three-dimensional systems. Thus, it is evident that higher-order cycloaddition reactions represent a very valuable tool for the construction of unusual heterocyclic frameworks. It is anticipated that the present review will capture the imagination of organic chemists and will stimulate further investigations in this area.

References

1. Ciabattoni J, Anderson HW (1967) Tetrahedron Lett 8:3377
2. Okamoto J, Yamabe S, Minato T, Hasegawa T, Machiguchi T (2005) Helv Chim Acta 88:1519
3. Paquette LA, Horton N (1968) Tetrahedron Lett 9:2289
4. Oda M, Funamizu M, Kitahara Y (1969) Chem Commun, p 737
5. Kende AS (1966) J Am Chem Soc 88:5026
6. Kende AS, Lancaster JE (1967) J Am Chem Soc 89:5283
7. Cantrell TS (1971) J Am Chem Soc 93:2540
8. Cantrell TS (1975) Tetrahedron Lett 16:907
9. Morita N, Kitahara Y, Asao T (1972) Tetrahedron Lett 13:869
10. Morita N, Kabuto C, Kitahara Y (1972) Tetrahedron Lett 13:4379
11. Truce WE, Lin C-IM (1973) J Am Chem Soc 95:4426
12. Ciabattoni J, Cabell M (1968) Tetrahedron Lett 9:2693
13. Maas G, Regitz M, Urgast K, Hufnagel M, Eckes H (1982) Chem Ber 115:669
14. Hayakawa K, Nishiyama H, Kanematsu K (1985) J Org Chem 50:512
15. Julino M, Bergstraesser U, Regitz M (1995) J Org Chem 60:5884
16. Saito K, Ando S, Kondo Y (2000) Heterocycles 53:2601
17. Machiguchi T, Hoshino M, Ebine S, Kitahara Y (1973) J Chem Soc Chem Commun, p 196

18. Machiguchi T, Otani H, Ishii Y, Hasegawa T (1987) Tetrahedron Lett 28:203
19. Machiguchi T (1995) Tetrahedron 51:1133
20. Machiguchi T, Hasegawa T, Kano Y (1993) Bull Chem Soc Jpn 66:3699
21. Machiguchi T, Kano Y, Hasegawa T (1990) Chem Lett, p 563
22. Machiguchi T, Hasegawa T, Itoh S, Mizuno H (1989) J Am Chem Soc 111:1920
23. Machiguchi T, Hasegawa T, Otani H, Ishii Y (1987) J Chem Soc, Chem Commun, p 1375
24. Machiguchi T, Yamabe S (1990) Chem Lett, p 1511
25. Cabrino R, Biggi G, Pietra F (1974) Synthesis 276
26. Machiguchi T, Yamabe S (1990) Tetrahedron Lett 31:4169
27. Yamabe S, Kawajiri S, Minato T, Machiguchi T (1993) J Org Chem 58:1122
28. Machiguchi T, Hasegawa T, Ishii Y, Yamabe S, Minato T (1993) J Am Chem Soc 115:11536
29. Machiguchi T, Hasegawa T, Otani H, Yamabe S, Mizuno H (1994) J Am Chem Soc 116:407
30. Machiguchi T, Minoura M, Yamabe S, Minato T (1995) Chem Lett, p 103
31. Minato T, Yamabe S, Ishiwata A, Hasegawa T, Machiguchi T (1999) J Mol Str (Theochem) 461–462:359
32. Schmack W, Nakazawa T, Hafner K (2000) Tetrahedron Lett 41:8255
33. Nair V, Abhilash KG, Suresh E (2006) Tetrahedron Lett 47:9329
34. Yamamoto K, Kajigaeshi S, Kanemasa S (1977) Chem Lett, p 85
35. Truce WE, Shepherd JP (1977) J Am Chem Soc 99:6453
36. Yamamoto K, Kajigaeshi S, Kanemasa S (1977) Chem Lett, p 91
37. Takaoka K, Aoyama T, Shioiri T (2001) Heterocycles 54:209
38. Sanechika K-I, Kajigaeshi S, Kanemasa S (1977) Chem Lett, p 861
39. Gandolfi R, Toma L (1980) Tetrahedron 36:935
40. Hayakawa K, Nishiyama H, Kanematsu K (1985) J Org Chem 50:512
41. Ito K, Saito K, Takahashi K (1991) Heterocycles 32:1117
42. Ito K, Hara Y, Sakakibara R, Saito K (1995) Heterocycles 41:1675
43. Saito Y, Tomita M, Taniguchi H, Okabayashi H, Saito K (1995) Heterocycles 41:2181
44. Gamba A, Ganndolfi R, Oberti R, Sardone R (1993) Tetrahedron 49:6331
45. Nishiwaki T, Abe N (1981) Heterocycles 15:547
46. Hafner K, Kläs H-G, Böhm MC (1980) Tetrahedron Lett 21:41
47. Nitta M, Kawaji H, Kanomata N (1992) Tetrahedron Lett 33:251
48. Nitta M, Iino Y, Mori S, Takayasu T (1995) J Chem Soc Perkin Trans 1 1001
49. Nishiwaki T, Fukui T (1986) J Chem Res (S) 288
50. Yasunami M, Miyoshi S, Kanegae N, Takase K (1966) Bull Chem Soc Jpn 66:892
51. Yang PW, Yasunami M, Takase K (1971) Tetrahedron Lett 45:4275
52. Chen A, Yasunami M, Takase K (1974) Tetrahedron Lett 30:2581
53. Yasunami M, Chen A, Yang PW, Takase K (1980) Chem Lett, p 579
54. Yasunami M, Yang PW, Kondo Y, Noro Y, Takase K (1980) Chem Lett, p 167
55. Yasunami M, Chen A, Noro Y, Takase K (1981) Chem Lett, p 555
56. Kuroda S, Hirooka S, Iwaki H, Ikeda M, Nakao T, Obisu M, Yasunami M, Takase K (1986) Chem Lett, p 2039
57. Kuroda S, Yazaki J, Meda S, Yamazaki K, Yamada M, Shimao I, Yasunami M (1992) Tetrahedron Lett 33:2825
58. Kitamori Y, Yasunami M, Kabuto C, Takase K (1993) Bull Chem Soc Jpn 66:245
59. Yasunami M, Hioki T, Kitamori Y, Kikuchi I, Takase K (1993) Bull Chem Soc Jpn 66:2273
60. Nitta M, Nishimura K, Iino Y (1993) Tetrahedron Lett 34:2157

61. Kuroda S, Mouri M, Hayashi K, Oda M, Yamada M, Shimao I, Osaki H, Yasunami M (1994) Chem Lett, p 85
62. Ito K, Kawaji H, Nitta M (1994) Tetrahedron Lett 35:2561
63. Takayasu T, Nitta MJ (1999) J Chem Soc Perkin Trans 1 687
64. Liu C-Y, Chen C-C, Shie Y-J, Chung L-W, Cheng T-S, Shie M-Y, Lin S-Y, Tsai Y-L (2003) Tetrahedron 59:6341
65. Nair V, Abhilash KG (2006) Tetrahedron Lett 47:8707
66. Luo Y, Herndon JW, Cervantes-Lee F (2003) J Am Chem Soc 125:12720
67. Zhang L, Wang Y, Buckingham C, Herndon JW (2005) Org Lett 7:1665
68. Godfrey JC (1959) J Org Chem 24:581
69. Galbraith A, Small T, Boekelheide V (1959) J Am Chem Soc 81:582
70. Galbraith A, Small T, Barnes RA, Boekelheide V (1961) J Am Chem Soc 83:453
71. Boekelheide V, Miller A (1961) J Org Chem 26:431
72. Boekelheide V, Kertelj SS (1963) J Org Chem 28:3212
73. Uchida T, Matsumoto K (1980) Chem Lett, p 149
74. Ikeda S, Kajigaeshi S, Kanemasa S (1976) Chem Lett, p 367
75. Yerxa BR, Moore HW (1992) Tetrahedron Lett 33:7811
76. Simonyan VV, Zinin AI, Babaev EV, Jug K (1998) J Phys Org Chem 11:201
77. Iida H, Uekusa H, Ohashi Y, Hamana H, Machiguchi T, Matsumoto K (2004) Heterocycles 64:447
78. Pereira DE, Leonard NJ (1986) Tetrahedron Lett 27:4129
79. Meier H, Saul K, Jacob D (1993) Liebigs Ann Chem 313
80. Meier H, Mayer A (1994) Angew Chem Int Ed Engl 33:465
81. Meier H, Rose B, Schollmeyer D (1997) Liebigs Ann Chem 1173

Author Index Volumes 1–13

The volume numbers are printed in italics

Subject Index

The manufacturer's authorised representative in the EU is Springer
Nature Customer Service Centre GmbH, Europaplatz 3, 69115 Heidelberg,
Germany. If you have any concerns regarding our products, please
contact ProductSafety@springernature.com

Printed and bound by CPI Group (UK) Ltd, Croydon, CR0 4YY
01/10/2026
02225951-0009